BRITISH OIL POLICY
1919–1939

BRITISH OIL POLICY
1919–1939

B.S. McBeth

Routledge
Taylor & Francis Group

LONDON AND NEW YORK

First published 1985 in Great Britain by
FRANK CASS & CO. LTD.
2 Park Square, Milton Park, Abingdon, Oxon OX14 4RN

and in the United States of America by
FRANK CASS & CO. LTD.
711 Third Avenue, New York, NY 10017

Routledge is an imprint of the Taylor & Francis Group, an informa business

First issued in paperback 2016

British Library Cataloguing in Publication Data

McBeth, B.S.
    British oil policy 1919-1939.
    1. Petroleum industry and trade—Government
    policy—Great Britain
    I. Title
    333.8'232   HD9571.6

ISBN13: 978-0-7146-3229-2 (hbk)
ISBN13: 978-1-138-98796-8 (pbk)

Typeset by Essex Photo Set, Rayleigh.
Printed and bound in Great Britain by
A. Wheaton & Co. Ltd., Exeter.

TO
MARIA CRISTINA

CONTENTS

LIST OF ILLUSTRATIONS

Between pages 30 and 31

1. William Knox D'Arcy – c.1907. A wealthy Englishman who made his fortune from gold mining in Australia, and obtained a 60-year-concession to search for oil in Persia.

2. Persia – H.I.M. Nasir al-Din Shah. He granted several concessions which included oil rights before his successor, Muzaffar al-Din, signed the D'Arcy Concession on 28th May 1901.

3. Mr C. Greenway at Sar-i-Pul, on his way from Chiah Surkh to Tehran – 1911.

4. Sir Robert Waley Cohen.

5. Sir Henri Deterding (1865–1939).

6. Marcus Samuel, First Viscount Bearsted (1853–1927), joint founder of the Shell Transport and Trading Company.

7. Sir John Cadman seen before leaving Tehran airport at the conclusion of Concession negotiations – 1933.

8. Mr. Reynolds, Mr. Willans and Mr. Crush having lunch, Persia c.1910. D'Arcy employed George Reynolds, (far left), to conduct drilling operations. Funding was assumed by the Burmah Oil Company in 1905, and Burmah founded Anglo-Persian in 1909.

9. Early geological survey party in Persia.

10. 'Colonel' Edwin L. Drake talking with Peter Wilson (left), a Titusville, Pennsylvania, druggist. On the extreme right in the background is 'Uncle Bill' Smith, Drake's head driller. This is where the modern oil industry started.

11. D'Arcy's first well – Chiah Surkh. Chiah Surkh was the scene of the first drilling in 1902. Oil was eventually struck in commercial quantities at Masjid-i-Sulaiman on 26th May, 1908.

12. No. 1 discovery well, Masjid-i-Sulaiman, Persia. A 'gusher' with wooden derrick typical of the period when precautions to control the flow of oil were not always adequate.

Nos. 1–3, 7–9, 11–13, 19, 22 by permission of British Petroleum; nos. 4–6, 10, 14–18, 20, 21 by permision of Shell International Petroleum Co. Ltd.

LIST OF TABLES

xi

ABBREVIATIONS

ADM	Admiralty
Anglo-American	Anglo-American Oil Co.
Anglo-Persian	Anglo-Persian Oil Co.
Anglo-Saxon	Anglo-Saxon Petroleum Co.
Asiatic	Asiatic Petroleum Co.
BOD	British Oil Developments Co.
BT	Board of Trade
Burmah	Burmah Oil Co.
CAB	Cabinet Office
Exxon	Standard Oil Co. (New Jersey)
FO	Foreign Office
Gulf	Gulf Oil Corp.
IPC	Iraqi Petroleum Co.
MGO	Mene Grande Oil Co.
MUN	Ministry of Munitions
Pearson	S. Pearson & Sons Ltd.
POWE	Ministry of Power
Shell	Royal Dutch-Shell Group
SOV	Standard Oil Co. (Venezuela)
TPC	Turkish Petroleum Co.
USGPO	United States Government Publications Office.

ACKNOWLEDGEMENTS

I am most grateful to my wife, María Cristina, and parents for their assistance in the preparation of this work. I owe much to E.F. Jackson for his detailed and meticulous comments on the manuscript and for his general encouragement. I am also indebted to Malcolm Deas and Robert Mabro for having read an earlier draft. Part of the research for this work was financed by a Foreign Area Fellowship granted by the Social Science Research Council (U.S.) and the American Council of Learned Societies. For any error of omission or commission I alone am accountable.

Brian McBeth

Introduction

In the early 1900s, especially after the invention of the Diesel engine, world oil consumption increased enormously, and the industrialised countries realised that oil was going to play an important role in the twentieth century. For strategic reasons the Admiralties of both Britain and the U.S. were at the forefront in stimulating the change-over from coal to fuel oil by the policy decision to change from coal-fuelled vessels to oil-powered vessels.[1] Indeed, in 1913 the British Admiralty under the First Lord, Winston Churchill, felt that the only way to safeguard its oil supplies was by the British government acquiring a majority stake in a British oil company. Consequently, the following year the British government acquired 51 per cent of *Anglo-Persian Oil Co.*'s[2] stock. *Anglo-Persian* held the oil rights over most of Persia and half the shares of the *Turkish Petroleum Co.*[3], and *TPC* held the oil rights over the Turkish Vilayets of Mosul and Baghdad (now Iraq).[4] The agreement represented the first open acknowledgement that the British Empire so long dependent on its own coal reserves found itself depending on foreign, mainly American, oil sources. It was during World War I however that the industrialised world first became acutely aware of the importance of oil. Although the U.S. produced 68 per cent of the world's total oil production the added burden of supplying the Allies with oil during the war and the enormous increase in industrial consumption[5] depleted its oil stocks, and there were strong fears that the country was running out of oil. The scarcity of oil experienced during the war also demonstrated to the British government the stranglehold which *Standard Oil Co. (New Jersey)*[6] and the *Royal Dutch-Shell Group*[7] (the duopoly which controlled the international oil market) held over British oil supplies. But more importantly, over two-thirds of British supplies came directly from U.S. oilfields, placing the British Empire at the mercy of the vicissitudes of American oil production. By the end of the war it became apparent that both the U.S. and Britain (for different reasons) would seek and stimulate the development of new sources of oil. During the initial post-war years attention centred on the Middle East and Persia because of Britain's traditional links with the area, but owing to political and logistic problems encountered during most of the 1920s[8] attention shifted to Venezuela.

On the surface Venezuelan production appeared to lessen Britain's dependence on American oil, but the two companies which supplied most of Britain's oil needs, *viz Shell* and *Exxon,* also controlled 90 per cent of Venezuela's production. Moreover, Venezuela's development as a major oil producer owed much to events which occurred in the American oil industry and to the changing structure of the world oil industry. *Anglo-Persian's* own position, supplying 20 per cent of Britain's needs, was undermined when in 1928 together with *Shell* and *Exxon* it signed the 'Pool Division Agreement'[9] to divide the world's oil markets among themselves. Britain then, in effect, had not secured an alternative source of oil supplies away from her traditional suppliers, and was about to enter the major armed conflict of this century with its oil supplies controlled by *Shell* and *Exxon.* These issues will be examined in more detail together with the various policies pursued by the British government to lessen its dependence on American oil supplies.

NOTES

1. cf. Miriam Jack, 'The purchase of the British Government's shares in the British Petroleum Company, 1912-1914', *Past and Present* No.39 (1968), 139-69; and, Chester Lloyd-Jones, 'Oil in the Caribbean and elsewhere', *North American Review* 202 (Oct. 1915), 536-43.
2. Hereinafter *Anglo-Persian.*
3. Hereinafter *TPC.*
4. cf. 'Agreement with Anglo-Persian Oil Company, with an Explanatory Memorandum and the report of the Commission of Experts on their local investigation' (Cmd.7419) *PP* LIV(1914), 505-16.
5. For example, the number of cars increased between 1914-18 by 250 per cent from 1.6 million to 5.6 million, and the number of trucks during the same period increased by 518 per cent from 85,000 to 525,000 (cf. John W. Frey & H. Chandler Ide, *A History of the Petroleum Administration for War, 1941-5* (Washington, 1946).
6. Hereinafter *Exxon.*
7. Hereinafter *Shell.*
8. cf. E.H. (Nicholas) Davenport & S.R. Cooke, *The Oil Trusts and Anglo-American relations* (London: Macmillan & Co., 1923); Louis Fischer, *Oil Imperialism* (London: George Allen & Unwin, 1926); George S. Gibb & E.H. Knowlton, *The Resurgent Years, 1911-1927* (New York: Harper & Bros., 1956); Gerald D. Nash, *The United States Oil Policy, 1890-1964* (Pittsburgh: University of Pittsburgh,

1968); Gholan Reza Nikpay, 'The political aspects of foreign oil interests in Iran down to 1947' (Ph.D. Diss., The University of London, 1956); and Benjamin Shwadram, *The Middle East Oil and the Great Powers* (New York:Council for Middle Eastern Affairs, 1959) 2 ed.

9. Better known by the incorrect title of 'As Is' or 'Achnacarry Agreement'.

1

Background

At the turn of the century world oil consumption increased enormously. In Britain, for example, consumption per head of population increased from 7.4 gallons in 1901 to 15.5 gallons in 1914.[1] For strategic reasons the Admiralty was at the forefront in formulating Britain's oil policy. Admiral Fisher, First Sea Lord from 1904 to January 1910, who believed that war with Germany was inevitable, pursued a vigorous modernisation of the navy by replacing coal-burning ships with oil-fuelled vessels. He reasoned that the Admiralty would be better able to defend Britain with faster and more efficient oil-powered ships.

Similarly the American Navy was also at the forefront leading the change-over from coal-burning engines to fuel oil. During the Spanish-American war it received $15,000 to study the feasibility of adapting its ships to fuel oil. Four years later the investigation broadened when Congress granted $20,000 to ascertain the value of crude petroleum for naval uses. By 1907 all American battleships being built in the U.S. were powered by fuel oil. The dreadnoughts USS *Oklahoma* and *Nevada* ran exclusively on oil, and so did 41 destroyers. In 1915 Secretary of Navy Josephus Daniels declared that 'hence-forth all the fighting ships which are added to the fleet will use oil, the transition from coal to oil will mark an era in our naval development almost comparable with the change from black powder to smokeless powder for our guns'.[2]

Britain's general policy on oil was laid down in 1904 when it was decided that oil concessions on Crown lands in British colonies and India would only be granted to companies under British control. The reasons for adopting such a policy were both strategic and economic, in order to prevent foreign companies from holding back the development of the oilbearing lands. The policy was adopted at the insistence of the Admiralty with a view to securing supplies of fuel oil for the Navy, and was 'introduced at a time when the Standard Oil Company was predominant and anxious to find outlets for American oil'.[3] It was feared that *Exxon* would acquire and keep the concessions unworked in order to stifle competition with

American domestic supplies; for example, in India it was felt that by securing control of the Burmese oilfields the American company would keep the Indian market open for American kerosene.[4]

According to the Colonial Office, British control did not exclude foreign capital but the 'Crown, as landlord of the Crown land, exercises a landlord's right in preferring its own nationals as tenants'.[5] The Crown did not lease land to individual foreigners:

> but when it is a case of a corporation, the Crown insists on the corporation being registered in the British dominions and having its head office on British soil, on its having a British manager and a majority of British directors, and a majority of its shareholders being British subjects. Foreign capital is not excluded, only foreign control.[6]

Under this provision, British control was exercised in Australia (North Territory), India, Burma, Trinidad, British Guiana, British Honduras, Nigeria, Nyasaland, Gold Coast Colony, Brunei, Papua, Fiji, and Cayman Islands. Foreigners could hold concessions for prospecting on Crown lands in the UK, Canada, Australia, New Zealand, South Africa, Kenya, Zanzibar, Johore, Jamaica, Barbados, Sarawak, Somaliland, British North Borneo, and Federated Malay States. However, the policy produced disappointing results as few British oil companies were formed to develop the possible oil lands of Empire. In 1908 Henry wrote that:

> What is wanted in the future is the enterprising employment of British capital, labour, and skill in the numerous unproven territories in different parts of the Empire. No time should be lost in developing these Imperial resources because there is much truth in the statement that the oil era having dawned, the Admiralty has every reason to complain that British enterprise has done little to pave the way towards the satisfactory solution to the problems of a truly Imperial liquid fuel-supply.[7]

The international oil business and British oil supplies were controlled by the two largest oil companies in the world, *viz Exxon* and *Shell*. Up to 1911 when the Supreme Court of the U.S. dissolved *Exxon*, the American oil industry was dominated by that company. It maintained strict control of the industry through its ownership of the pipelines. It also used all possible means to prevent other companies from laying pipelines, such as buying up the land along the route of rival pipelines, enlisting the support of railways in refusing rivals the right of way across the rail tracks, obtaining control of independent pipelines, and by the payment of premiums on crude oil to producers in the immediate vicinity of an independent pipeline. *Exxon* would offer 5 to 10 cents per barrel above market price so as to cut off supplies to independent pipelines.[8] In 1890 the company expanded to Europe establishing the *Compagnie Générale des Pétroles* in France, the *Anglo-American Oil Co.* in the UK, the *Deutsche Amerikanische Petroleum Gesellschaft* in Germany, and various other companies.[9] In 1906 the Bureau of Corporations of the Department of Commerce found that *Exxon* controlled in the U.S. 84.2 per cent of crude oil consumed, and 86.5 per cent of refined products. The company also handled 87 per cent of total U.S. exports and 88.7 per cent of the total domestic oil marketed. In 1909, the U.S. Circuit Court for the Eastern Division of the Eastern Judicial District of Missouri found that *Exxon* had fomented a 'combination or conspiracy in restraint of trade and commerce in petroleum and its production among the mineral States in the Territories and with foreign nations'.[10] The Court decreed that *Exxon* should desist 'from exercising or attempting to exercise control, direction, supervision or influence over the acts of these subsidiary companies'.[11] On 15 May 1911, the Supreme Court confirmed this decree and ordered the company to dissolve its links with its subsidiaries. Three years later, in 1914, fearful of the zealous scrutiny of U.S. government agencies, *Exxon* centralised its foreign operations in London. The year 1911 is also a further watershed for the industry, for a fundamental change in the character of the industry occurred. From mainly producing kerosene, the industry increased its products range, of which the most important were fuel oil and petrol. Owing to the structure of the American oil industry crude oil production fluctuated so much that there were periods of great abundance and scarcity with wide price fluctuations. This destabilising element would be brought under control during the inter-war period with the introduction of prorationing measures, which had the effect of stabilising prices at relatively high levels, much to the benefit of the integrated oil companies such as *Exxon* and *Shell*.

Shell was registered in London in 1897 with an authorised capital of £42 million. In June 1903 it combined with *Royal Dutch Co.* and the Rothschilds' Russian interests to form a marketing organisation, the *Asiatic Petroleum Co.* Each company, however, retained its separate corporate identity.[12] In 1907 a more permanent combination was achieved when *Shell* merged with the *Royal Dutch Co.* Two new companies had to be formed, the *Anglo-Saxon Petroleum Co.*,[13] registered in the UK with a capital of £25 million, and responsible for marketing operations of the Group; and the *Bataafsche Petroleum Maatschappij*, registered in the Netherlands with a capital of Fl. 300 million, and in charge of crude oil production. *Shell* held 40 per cent of the stock of both companies and the *Royal Dutch Co.* the remaining 60 per cent. The development of *Shell* into a major oil company was achieved as a direct result of circumstances at the time, and by the financial acumen and policy objectives pursued by Henri Deterding. American oil companies had necessarily followed an insular pattern: after all 68 per cent of the world's oil was produced there and the U.S. was the largest consumer of oil. On the other hand, Holland and Britain had to import their oil from the four corners of the world, with the U.S. supplying the greater part of their needs. Deterding, whom Lord Fisher described as a 'Napoleon in boldness and Cromwellian in depth',[14] was responsible for this policy. Deterding explains:

> In the years of clash and conflict which followed from 1903, it used to be said that I had taken as our motto: 'Our field is the world.' And although I never chose this motto deliberately, the fact remains that, co-ordinated as we now are into a vast cosmopolitan undertaking – representing Great Britain through Shell; Russia and France through the Rothschilds – we had no other alternative but to expand, expand, expand. Had we restricted our trading purely within certain areas, our competitors could easily have smashed us by relying on the profit they were making in other countries to undercut us in price. So, to hold our own, we had to invade other countries, too.[15]

In 1876 the Nobel Brothers started producing oil in Russia. By 1913 their company had increased its capital to 30 million roubles and produced 14 per cent of Russia's output.[16] According to Liefman this group, together with that of Rothschild in France, in the 1890's formed the first international cartel for European oil sales. Later other Russian producers, together with the *Royal Dutch Co.* and *Nobel Rothschild*, combined with the *Deutsche Bank* to form in 1904 the *European Petroleum Union* to fight *Exxon*. This combine handled Russian, Galician and Romanian oil.[17] After a short period of intense fighting the Anglo – Dutch – German – Russian interests came to an arrangement with *Exxon* concerning the European market for illuminating oil.[18] In a later agreement *Shell* secured the monopoly rights to supply benzine to the whole of Europe. In 1910, Deterding and *Exxon's* European Director, Walter C. Teagle travelled to New York to reach an agreement with *Exxon* to stop competition. John D. Rockefeller, head of *Exxon,* however, did not accept Deterding's terms and price wars broke out in various markets. A fierce battle began in the Far East between the two rival companies. As a result of this competition *Shell* experienced a loss in profits as *Exxon* was 'trying in an indirect way to secure oilfields in the Dutch East Indies'.[19] *Exxon* was undermining *Shell's* position in their traditional markets by reducing their prices 'most and quickest in Netherlands, India, whilst in Holland, where we sell no kerosene, but only benzine, the benzine prices were reduced most'.[20] Deterding believed that these upheavals were the result of the obstinate desire of *Exxon* to control the entire world oil market. Deterding was therefore aware that, unless he competed with *Exxon* in the U.S., any losses *Exxon* made in their war against *Shell* could be offset by profits from their U.S. market. Thus in 1911, under cover created by the Supreme Court's dissolution decree of *Exxon,* Deterding started to lay plans to 'commence both producing and marketing operations in the United States'.[21] By the end of the First World War *Shell* had become a fully integrated oil company that could compete with *Exxon* on equal terms.

Britain's efforts to lessen its dependence on these two giants of the oil industry rested on two obscure British companies, viz the *Anglo-Persian* and *TPC.* On July 25, 1872 Shah Nasr-ed-Din granted the first major concession in Persia to exploit mineral resources to a British Baron Julius de Reuter. Its duration was for 70 years and covered the whole of Persia. The monopoly gave Baron de Reuter the right to construct railways and street car lines, and the exclusive right to exploit all the mineral resources of the country. According to Shwadran the concession was cancelled the following year after the Russians made clear to the Shah their displeasure with it during his visit to the St. Petersburg Court.[22] In 1889, after British diplomatic pressure was exerted, Baron de Reuter acquired a new concession which gave him the right to open a bank and exploit the mineral resources of the country, including oil, for 60 years. The Persian government would receive a 16 per cent royalty of the net profits. As a result, the *Imperial Bank of Persia* was formed, and in 1890 the *Persian Bank Mining Rights Corp.* was organised to work the mineral deposits. Exploration work carried out during 1891–3 by Jacques de Morgan, a French geologist, confirmed the existence of oil in the Mesopotamian–Persian border region. However, in 1899 the Persian Government rescinded the minerals concession alleging a lack of adequate development of the property. At the Paris Exposition of 1900 Edward Cotte, one time Secretary to Baron de Reuter, drew de Morgan's findings to the attention of General Kitabji Khan, Persia's Commissioner General. Kitabji tried unsuccessfully to persuade French capitalists to invest in the country. He then turned to Sir Henry Drummond Wolff, former British Minister at Teheran, to influence British capitalists to invest. Sir Henry convinced William Knox D'Arcy of the desirability of acquiring the property. D'Arcy sent out the geologist H.T. Burls to ascertain the possibilities of oil. His favourable report convinced D'Arcy upon his return to London, who dispatched Alfred L. Marriot, together with General Kitabji and Cotte, to acquire a concession in his name. Sir Henry Drummond Wolff used his influence with Sir Arthur Hardinge, his successor at Teheran, and with the Persian Government, by assigning shares in the proposed company to some of the most influential Persian ministers, including the Grand Vizier himself, so as to ensure that the concession was granted. The Russians once again objected to such a concession, and in order to placate them, the five major northern provinces (Azerbaijan, Gilan, Mazanderab, Khorasan, Astrabad) were left out of the concession. On 28 May 1901, Alfred Marriot, D'Arcy's attorney, and Shah Mozaffar-ed-Din, signed a 60-year concession with the Persian Government to receive a 16 per cent royalty on net profits. Two years later D'Arcy organized the *First Exploration Co.* with an authorized capital of 600,000 £1 shares, of which 544,000 were issued and fully paid, and subscribed by D'Arcy.[23] However, D'Arcy ran out of money and started to look for further financing. According to Shwadran because of its fear that the concession might fall into American or Dutch hands the British Admiralty asked D'Arcy to postpone any negotiations with foreign buyers until British interests could be found to take over the concession. In 1905 E.G. Pretyman, Civil Lord of the Admiralty persuaded the Scottish-Canadian financier Lord Strathcona[24] of *Burmah*[25] to take over the concession. In May of the same year the *Concessions Syndicate Ltd.*, with a capital of £100,000 and with D'Arcy as a Director, was formed to exploit the Persian concession. In April 1909, Lord Strathcona and *Burmah*, the main shareholders in the *Concessions Syndicate*, transferred their holdings to the *Anglo-Persian*, a company registered in London on 14 April 1909, with an initial capital of £2 million, having Lord Strathcona as its Chairman and D'Arcy as its managing Director.[26]

Two important political events took place in Persia at the time. First, constitutional government came to the country, and, second, under the Anglo-Persian Agreement of 1907, the country was divided into three spheres of interest: the northern part reserved for Russia, the southern for Britain and a neutral central zone. One result for Persia was a 'further entrenchment of foreign powers on her soil and a further weakening of the control of the Teheran Government over the zones under the influence of Russia and Great Britain'.[27] Another was that prospecting parties would have to seek agreement not only with the regional government but also with the local people.[28] For example, when the company struck oil at Maidan-i-Naftum, an area controlled by the Bakhtiari Khans, in order to obtain their co-operation an agreement was signed granting them a 3 per cent equity share in the *Bakhtiari Oil Co. Ltd.,* a company formed in 1909 to exploit the oil in their territory.[29] Following negotiations with Sheikh Khazaal, who controlled the territory of Muhammarah, the Abadan refinery was constructed in 1912. The Sheikh agreed to provide protection for the pipeline at the company's expense, and to accept in advance ten years' rent of £650 per annum.

The early history of *TPC* is just as varied. In 1908, Admiral Colby M. Chester, the same skilful negotiator used to settle the claims for compensation for life or property suffered by American missionaries in the Armenian Massacres, returned to Constantinople with the purpose of obtaining railway and mining concessions. According to Shwadran he had the backing of the New York Chamber of Commerce and the New York State Board of Trade, and was 'supported by President Theodore Roosevelt and Secretary of State Elihu Root'.[30] The following year he obtained contracts to construct a port and three railway lines which would pass through the Vilayets of Mosul and Baghdad, the rich oilbearing lands. The railway contracts gave Chester the mineral rights to be found within 20 kilometres of the railway track. On his return to the U.S. Chester organised the *Ottoman-American Development Co.* under the laws of New Jersey with a capital of $600,000. On 9 March 1910, contracts were signed by the Turkish Public Works Minister and sent to the Grand Vizier and Parliament for ratification the following year. But the Turco–Italian and Balkan wars as well as the First World War prevented this.

The Germans had also been active in the area. In 1904 they had acquired from Sultan Abdul Hamid a railway concession which putatively gave them the right to survey for oil in the Mosul and Baghdad Vilayets. For this they organized the *Anatolian & Baghdad Railway Concession* in which Sultan Abdul held shares. The Young Turks deposed the Sultan before work could start and the concession was subsequently cancelled. The *Turkish Petroleum Co.* in which *Shell*, the *National Bank of Turkey* and the *Deutsche Bank* were partners, was organized in 1912 to revive the 1904 claim,[31] but the Turkish government refused to recognise the company's claim. D'Arcy too had tried to acquire the mineral rights to the two Vilayets. He had been promised by successive Turkish Grand Viziers that he would be awarded the oil rights to the two Vilayets but was unable to secure written confirmation. A compromise was reached between *TPC* and D'Arcy when it was seen that neither was going to obtain the sole mineral rights to the area. *Anglo-Persian* would secure British government assistance to persuade the Turks to ratify the 1904 concession. The British government agreed to this because it feared that *Shell* would take over *Anglo-Persian*.

In 1913 there was strong pressure on *Anglo-Persian* to merge with *Shell*. The reason for the former's resistance was the promise that it would secure the Mesopotamian concessions. *Anglo-Persian* decided to seek greater backing from the British government for its enterprise. *Anglo-Persian* also felt that the failure of *Shell* to acquire the company would introduce direct competition between the two companies, and if *Shell* acquired the *TPC* concession it would start a 'war of rates in the Middle East market and thus force *Anglo-Persian* to merge with it that way'.[32] Sir Charles Greenway, *Anglo-Persian's* Chairman, declared that once *Shell* achieved this they would force the price of oil up by developing the large Persian oilfields at a slow rate. Sir Charles further argued that if *Anglo-Persian* took over *TPC* then the Royal Navy would be assured of a large source of oil. It was of paramount necessity to keep *Anglo-Persian* British in order to provide cheap oil for the Navy and support Imperial interests in an area vital to the Indian subcontinent. It was felt that:

> To allow foreign interests, which were *ipso facto* unreliable in times of national stress, to become established next door to the young Persian oil industry, purely British in control and sentiments, would be tantamount to aiding the potential destruction of British naval supremacy.[33]

The British and German governments then undertook to induce the

Turkish government to recognise the 1904 claim. Turkey and Persia had had a long history of border disputes. These were settled when the Turco-Persian Protocol of 7 November 1913 was signed by the two countries. Under the Protocol a narrow strip of land, the Qasr-i-Shirin area (covered by the D'Arcy concession) was awarded to Persia, while the Khanagin area, where *Anglo-Persian* had producing oil wells, was transferred to Turkey. The British government obtained an undertaking from Turkey that the D'Arcy concession would be recognised in the transferred territories. Turkey also granted *Anglo-Persian* the privilege of constructing pipelines through Turkish territory to the sea. At the same time *TPC's* 1904 claim was recognised. In 1914 *Anglo-Persian* agreed to surrender its claim in return for a share in *TPC's* equity. A reconstituted company was formed in March that year in which *Anglo-Persian* acquired a 50 per cent interest, and the remaining shares held by the *National Bank of Turkey* were divided between the *Deutsche Bank* and the *Anglo-Saxon* giving each of them a 25 per cent stake in the new company.[34]

Owing to a general lack of British commercial interests to develop the Empire's sources of oil, and the need to secure permanent oil supplies, it became apparent to the Admiralty that it would be of vital strategic importance for the British government to acquire at least a majority stake in a British oil company which held promising oil bearing concessions.

The Admiralty had begun trials with oil-burning warships in 1901. A committee appointed by the Admiralty in 1904 to look into the use of oil favoured the adoption of fuel-oil-powered engines in the Navy's ships. The most pressing problem that needed to be solved was the availability of supplies.[35] Nevertheless, the decision was taken to adopt oil as a fuel for vessels in the Navy and a number of destroyers were commissioned. The Admiralty, however, were thwarted in this by the Liberal government which came to power at the end of 1905, and it was not until the new Conservative government of November 1911 that the Admiralty managed to put its plan into effect. From now on it would actively pursue its policy to rely more on fuel oil than on coal, and as a first step five oil-powered battleships were commissioned.

The Admiralty was faced with two major problems; one was the price of oil which had increased considerably in 1911–12, and the other was the source of supply. *Exxon* was expanding fast, seeming to make Britain dependent on American oil, and *Shell* was looked at with distrust because of what was feared to be its domination by

Germany. The problem of supply became more acute as the possibility of war with Germany increased. In March 1912 Churchill, First Lord of the Admiralty, stated in the House of Commons that the provision of adequate fuel oil supplies was one of the most difficult problems confronted by the Admiralty.[36] Consequently, a Royal Commission on Fuel and Engines, under Lord Fisher, was appointed to report on 'the means of supply and storage of liquid fuel in peace and war and its applications to warship engines, whether directly or by internal combustion'.[37] Churchill in 1913 believed that the Scottish shale industry, which had been producing shale oil in commercial quantities since 1858, could produce between 400,000 and 500,000 tons of oil for the next 150 years.[38] A further problem was the high price of oil which Churchill felt was the direct result of 'vast and formidable schemes on the part of a comparatively small number of wealthy combinations to control the oil market and raise and maintain prices.[39] The Admiralty's dependence on oil was so great that Churchill argued it had to have guarantees of being 'able to buy a steady supply of oil at a steady price'.[40] By not controlling its oil supplies the Admiralty would be forced to purchase it at uneconomical prices. Lord Fisher's Commission on Fuel and Engines issued three reports (which remained confidential – a fact that caused a great deal of controversy)[41], all stressing the importance of making long-term contracts from a wide variety of sources and recommending that large reserves should be accumulated in the UK and the Empire.[42] In 1913 the Admiralty had 25 different contracts for fuel oil. Lord Fisher's Commission felt that the *Mexican Eagle Co.*, the largest Mexican oil producer and owned by Lord Cowdray, could supply the Admiralty's needs.

The Opposition severely criticised Churchill for building oil-powered ships without first obtaining a secure source of oil. Lord Beresford felt that shale oil was not a viable solution and that Mexican production was in jeopardy because of the political turmoil experienced by the country. In addition, Mexican oil was unsuitable for use in the ship's engines. Lewis Harcourt, Colonial Secretary, had hopes that Trinidad could supply the Admiralty's needs and had entered into an agreement with the *British West Indies Petroleum Syndicate*, a subsidiary of *Shell*.[43] The Admiralty also had a standing order with *Burmah* for 100,000 tons of fuel oil a year in war-time, and had in addition signed a large forward contract with *Anglo-Persian* in 1913. A Commission headed by Rear Admiral Sir Edmond J.W. Slade went to Persia in October 1913 to look at the resources of the company.[44]

In a debate in the House of Commons on the Navy Estimates for 1913-14, it was stated that it was desirable to keep 'alive independent competitive sources of supply, so as to safeguard the Admiralty from becoming dependent on any single combination'.[45] Moreover, in time of war 'if some areas should be closed others will remain open'.[46] However, it was also apparent that the Empire's total production of 2.5 per cent of the world's total was not sufficient to meet its needs. It thus became clear that 'while it is not desirable to draw all our supplies from one source it is essential that the fields over which His Majesty's Government will have control shall be so developed that in times of emergency they will be able to supply at short notice any deficiencies that may arise through the failure of deliveries elsewhere'.[47] The aim was for the Admiralty to become the independent owner and producer of its own supplies of oil.[48]

On the eve of World War I the Admiralty had 120 cruisers, 5 battleships and 16 light cruisers which burned oil, and which in the case of the cruisers needed 6,000 tons of oil a day to operate.[49] The price of oil was four times greater than the price of coal and the need to conserve oil stocks meant that the new torpedo flotilla had not been exercised. On 3 March 1914, F. Hall asked in the Commons whether there was any truth in reports that many destroyers had been temporarily laid up as a result of a shortage of fuel oil. Lord Beresford on 18 March replied that 62 new ships of the 'L' class of T.B.D. were due to be launched soon, but 'we have not been able to try them because we have not the oil'.[50] Churchill did not deny these allegations only stating that the Admiralty's oil reserves were being built up with oil being purchased in large quantities,[51] and requested from Parliament £500,000 in additional funds to increase oil reserves.[52] Churchill, however, intended taking a large stake in *Anglo-Persian*. The Admiralty's Commission on the Persian oilfields, which reported on 6 April 1914, gave the company a glowing report when it concluded that:

> We are satisfied that the Company's concession is a most valuable one, and providing no unforeseen factor intervenes, the existing field is capable, with proper development of supplying a large proportion of the requirements of the Admiralty for a considerable period, while the whole concession judiciously worked would probably safeguard the fuel supply of His Majesty's Navy.[53]

The Commission also felt that the company would not possibly develop such an extensive area on its own and therefore it would be desirable for the British government to enter into an agreement with the company as this would ensure efficient supervision of the capital invested, a proper conservation of the oilfields, and the general future development of the concession. The company gained also because the sales contracts it entered into with the armed services spared them the expense and risk of setting up marketing organisations to compete with the other large companies.[54] As a result on 20 May 1914 the British government entered into an agreement with the company whereby it subscribed to 51 per cent of the company's capital. There would be two ex-officio directors appointed by the Treasury, one of whom would represent the Admiralty directly, having the power of veto 'over all acts of the Board and Committees of the company and its subsidiaries'.[55] There was some opposition to the government but on 10 August 1914, the Anglo-Persian Oil Co. (Acquisition of Capital) Bill received the Royal Assent.[56] The agreement represented the first open acknowledgement that the British Empire, so long dependent on its own coal reserves, found itself depending on foreign, mainly American, oil sources. The experience during the First World War only served to reinforce this point more strongly.

NOTES

1. Royal Commission on the Natural Resources, Trade and Legislation of certain portions of His Majesty's Dominions, Royal Commission Dominions, 'Memorandum and tables relating to the Food and Raw material requirements of the United Kingdom', *PP* 1914–16, Vol. xiv, Cmd.8123 (1915), pp.371-498, Table 133, p.463.
2. Lloyd-Jones, *op. cit.*, p.537.
3. POWE 33/275 Petroleum Board, 'Oil Concessions in British Colonies and Protectorates. British control of companies', July 1929.
4. POWE 33/353 'Petroleum-Nationality. Restrictions in Oil Leases on Public Lands in British Territory – Statement of the present position by the Petroleum Department', 18 July, 1923.
5. POWE 33/353 Colonial Office, 'Memorandum. British Control of Oil Mining Companies', June 1923.
6. *Ibid.*
7. J.D. Henry, *Oil fuel and the Empire* (London:Bradbury, Agnew & Co., 1908), p.176.
8. FO 368/140 Des.90 Commercial James Bryce to Sir E. Grey, Intervale, 20.7.07.

9. Robert Liefman, 'International Cartels', *Harvard Business Review*, 5:2 (Jan. 1927), 129-48.
10. FO 371/12835 Petroleum Dept. 'Notes on the various Groups of Standard Oil Companies', 14.9.27.
11. *Ibid.*
12. Sir Henri Deterding, *An International Oilman* (London: Ivor Nicholson & Watson Ltd., 1934), pp.65-71.
13. Hereinafter *Anglo-Saxon.*
14. Pierre L'Espagnol de la Tramerye, *The World Struggle for Oil* (London:George Allen & Unwin, 1923), 3rd ed., p.67.
15. Sir Henri Deterding, *An International Oilman* (London:Ivor Nicholson & Watson Ltd., 1934), p.176.
16. cf. Anton Mohr, *The Oil War* (London:Martin Hopkinson & Co. Ltd., 1926).
17. G.S. Gibb & E.H. Knowlton, *The Resurgent Years, 1911-1927* (New York: Harper & Bros., 1956).
18. Liefman, *op. cit.*
19. Royal Dutch Co., *Annual Report, 1911*, p.10.
20. *Ibid.*, p.16.
21. Gibb & Knowlton, *op. cit.*
22. cf. Benjamin Shwadran, *The Middle East, Oil and the Great Powers* (New York: Council for Middle Eastern Affairs, 1959), 2nd ed.
23. *Skinner's Oil Manual, 1913*, p.56.
24. Created a Baron in 1897. High Commissioner for Canada, 1896-1911. *Who's Who, 1897-1915.*
25. Company registered in Edinburgh on 15 May 1902 and held a 39-year oil concession in Burma. *(Skinner's Oil Manual, 1912* p.34).
26. *Skinner's Oil Manual, 1913*, p.8-9.
27. Shwadran, *op. cit.*, p.20.
28. At the end of 1907 the Indian government sent Lt. Arnold T. Wilson with a detachment of soldiers to the neighbourhood of Shustan to protect the drillers from local attacks.
29. *Skinner's Oil Manual, 1913*, p.19.
30. Shwadran, *op. cit.*, p.197.
31. Sir Ernest Cassel was instrumental in this because he held shares in the National Bank of Turkey.
32. Marian Ruth Kent, 'British Government interests in the Middle East Oil Concessions, 1900-1925' (PhD.Diss., University of London, 1968), p.62.
33. *Ibid.*, p.62.
34. cf. Ludwell Denny, *America conquers Britain* (New York: Alfred A. Knopf, 1930), E.H. (Nicholas) Davenport & S.R. Cooke, *The Oil Trusts and Anglo-American Relations* (London: Macmillan & Co., 1923); and, D.J. Payton Smith, *Oil. A Study of War-time Policy and Administration* (London: HMSO, 1971), Appendix 1, 1914, Agreement on the Mesopotamian Concession, pp.26-7.
35. *Hansard*, 1904 Vol. 137, 'Navy Estimates Debate, 1904-5', 30.6.04.
36. *Hansard*, 1912 Vol. 35, Col.1560.
37. *Hansard*, 1912 Vol. 41, Col.2504.
38. The Admiralty placed an order of 30,000 tons before World War I.
39. *Hansard*, 1913 Vol. 55, Col.1473.
40. *Hansard*, 1913 Vol. 55, Col.1474.
41. *Hansard*, 1913 Vol. 55, Col. 1219 and 1913 Vol. 59, Col.1872.
42. Payton Smith, *op. cit.*, and 'Agreement with the Anglo-Persian Oil Company, with an explanatory Memorandum and the Report of the Commission of Experts on their local investigations', *PP* 1914, Vol. liv, Cmd. 7419, pp.505-39.

43. *Hansard,* 1913 Vol. 56, Col. 783.
44. 'Agreement with Anglo-Persian', *op.cit.,* 'Admiralty Commission on the Persian Oilfields', 6.4.14, pp.527-38.
45. *Ibid.,* p.509.
46. *Ibid.,* p.509.
47. *Ibid.*
48. *Ibid.,* 'Extract of speech of First Sea Lord of the Admiralty–Debate House of Commons on Navy Estimates for 1913-4', 17.7.13, pp.513-7.
49. *Hansard,* 1914 Vol. 59, Major Archer-Shee, 2.3.14.
50. *Hansard,* 1914 Vol. 59, 'Supply (Navy Estimates) Order for the Committee', 18.3.14, Col.2103.
51. *Hansard,* 1914 Vol. 59, 17.3.14.
52. *Hansard,* 1914 Vol. 59, 'Supply Navy Supplementary Estimates', 2.3.14.
53. 'Agreement with Anglo-Persian', *op. cit.,* p.532.
54. Sir Marcus Samuel, Chairman of *'Shell' Transport and Trading Co.* protested to Churchill about this agreement arguing that the British government was now competing against other commercial concerns with the 'unfair advantages of government financial and diplomatic support' (Kent, *op. cit.,* p.253). Sir Marcus further argued that *Anglo-Persian's* production would swamp the market and drive the price of oil down. He suggested to Sir F.J.S. Hopwood of the Admiralty that:

> To mitigate this danger I suggest that before increasing the quantities of liquid fuel deliverable by the Persian Company to the Government beyond an agreed minimum figure, the Government should give the option to all producers in the British Empire . . . to supply fuel to the Admiralty, provided they can do so at the same price delivered in the United Kingdom, and that only if these sources fail should the Admiralty have recourse to Persia for increased quantities. (CAB 37/120/68 Sir Marcus Samuel to Sir F.J.S. Hopwood, 28.5.14.)

Churchill felt that this suggestion 'to keep prices up to the black-mailing levels deserves notice'. (CAB 37/120/68 Letter initialled W.S.C., 9.6.14)
55. 'Agreement with Anglo-Persian', *op. cit.,* p.507.
56. *Hansard,* 1914 Vol. 64.

2

Britain's Efforts to Achieve Oil Independence

In 1916 two foreign controlled companies, the *Asiatic Petroleum Co.*,[1] part of *Shell,* and the *Anglo American Oil Co.*,[2] a subsidiary of *Exxon,* supplied over 90 per cent of the UK's petrol needs. The largest single supplier was the USA; for example in 1914 it accounted for 84 per cent, 80 per cent and 35 per cent of lubricating oil, lamp oil, and petroleum spirit respectively.[3] Production from the Scottish shale oil fields supplied 18.8 per cent of Britain's oil needs in 1901 declining to 10.8 per cent in 1914.[4]

In order to meet its army and navy contracts Waley Cohen, Managing Director of the *Asiatic,* warned the Board of Trade in 1916 that his company would have to restrict commercial sales of petrol.[5] With the restrictions Cohen anticipated a serious shortage of petrol in the UK in the immediate future. Similarly, the *Anglo-American* company, Britain's largest single importer of oil, informed the Board of Trade that petrol imports would be restricted because two of their tankers had been impounded. In June 1916 a petrol licence duty of 6d per gallon of petrol was introduced which was expected to raise nearly £1 million in revenue.[6]

Owing to the Empire's low production of oil it became clear to the Board of Trade and the Foreign Office that it was of the utmost necessity to secure control of foreign oil sources of supply by bringing about a combination of British interests operating in foreign countries.[7] The question had important future foreign policy implications for Britain in the Middle East. It was essential, the Foreign Office reasoned, that German pre-war economic penetration in the oilfields of Egypt and Mesopotamia should be taken over by a wholly controlled British group after cessation of hostilities. The role of *Anglo-Saxon* in the pre-war Mesopotamian negotiations for oil had caused a certain amount of embarrassment to the Foreign Office. A strong British company was therefore needed as it appeared that neither *Burmah* nor *Anglo-Persian* disposed of the 'economic independence, the areas of supply or the commercial ability to enable them to fulfil the necessary conditions'.[8] *Anglo-Persian* as an oil marketing company was not commercially viable because of the constraints imposed on its trading activities by *Shell's* possession of certain marketing rights in respect of various grades of oil. The only way that *Anglo-Persian* could survive was with the aid of government subsidies. The only viable short-run solution then was to bring *Shell* under British control, and the situation at the time appeared to the Foreign Office:

> opportune for such a development; the directors of the Shell Transport Company [sic] are, it is believed, favourable to some such scheme, and prepared to make sacrifices in order to enter into closer relations with His Majesty's Government, even agreeing to some system for the regulation of prices; the Directors of the Royal Dutch Company are also, so Sir E. Grey is informed, not adverse to adopting a like attitude.[9]

At the same time Lord Cowdray proposed the formation of an all-British oil company merging his Mexican interests with those of *Burmah* and *Anglo-Persian*.[10] A committee under Walter Runciman, President of the Board of Trade, was appointed to consider 'what measures are necessary to secure adequate supplies of petrol for the purpose of the war and other essential needs'.[11] It recommended that *Shell* and *Burmah* should amalgamate to form the *Imperial Oil Co.* with permanent British majority shareholding and control. The proposal was approved by the War Cabinet in principle, but the Admiralty fearing that its oil supplies would not be guaranteed found it unacceptable and negotiations collapsed.[12]

The matter, however, would not rest here and questions were asked in Parliament as to how the government would lessen its dependence of foreign oil supplies.[13] Adequate and steady oil supplies became one of the most pressing requirements the British government needed to resolve. It was felt that the outcome of the war, and indeed the future of the Empire rested on a continuous, ever increasing supply of oil. Walter Long, President of the Local Government Board, expressed the mood of the Cabinet when he stated in the House of Commons that 'you may have new men, munitions and money, but if you have not got oil all other advantages will be of comparatively little value'.[14] The dependence on U.S. supplies only served to heighten the desire to secure alternative sources of oil. For example, the final report of the Royal Commission on Natural Resources, Trade and Legislation in certain portions of the Dominions (set up to look at the UK's and the Empire's sources of raw materials), felt that the question of petroleum supplies to the Empire was 'so important as to deserve a special mention'.[15] It urged that:

> in view of the importance of petroleum as an asset to the Imperial and Dominion Navies, that where in any part of the Empire an extensive oil-bearing area is found, steps might be taken to reserve some portions from public competition, so that, where circumstances permit, special provisions may be made for their development and the employment of the product for naval purposes.[16]

Sir John Cadman, Professor of Geology at Birmingham University and attached to the Ministry of Munitions, recommended the setting up of a Petroleum Department.[17] Sir Marcus Samuel of *Shell* seized the opportunity to request British assistance in developing properties recently acquired in Venezuela. Sir Marcus had no doubt that Venezuela, if not equalling the U.S., would become a large supplier in the near future. However, the company was experiencing difficulties in transferring equipment and materials to Venezuela and Curaçao.[18] He therefore offered to place Venezuela's production at the 'disposal of the Admiralty, not requiring them to give us any financial assistance of *any kind* as a condition of a contract'[19] if 'the necessary assistance in procuring means of moving and refining the oil could be given to us as an emergency measures. I should not like to say what limit we would put on the oil which we could deliver within a measurable time, say six months'.[20] Walter Long took up Sir Marcus Samuel's proposal and instructed Lord Northcliffe, who headed the British Mission in the U.S., to support Waley Cohen's application to the newly created Inter-Allied Petroleum Council to obtain export licences for *Shell* to send oil equipment to its subsidiaries in the 'Dutch East Indies, Venezuela and Mexico'.[21]

In June 1917 the Admiralty approached Lord Cowdray with the view that he should start drilling for oil in the UK. There was no doubt in Lord Cowdray's mind that there was oil in the UK but the drawback was that English law allowed competitive drilling, and this would lead to uncontrolled drilling activity as had happened in the U.S.. Lord Cowdray was willing to place at the government's disposal his company and its geological staff for the duration of the war free of charge. He would also bear all drilling costs on government owned land and on private land if better terms than those afforded by the Defence of the Realm Act (which meant that all work performed by his company would revert back to the landlord) were offered. Lord Cowdray had spent several years negotiating leases with landlords and felt the best policy for stimulating oil exploration was to set up a 'system of National Drilling licences'.[22] Soon afterwards, in order to encourage home production of oil (production of shale oil had increased considerably)[23] and avoid the 'prodigious waste of capital'[24] and depletion of oil, the government introduced in August 1917 the Petroleum (Production) Bill which would give the government the exclusive right of exploring and developing the oil deposits found in the UK or of awarding leases for this purpose. A fixed royalty of nine pence per ton of oil extracted would be levied to form the Petroleum Royalty Fund which would provide transport facilities for oil and the necessary drilling equipment.[25] The Bill, however, was withdrawn on 14 January 1918, because of strong opposition to the proposed petroleum royalty by the country's landed gentry.[26]

Nevertheless, Britain's dependence on foreign oil companies did not diminish and the problem of creating a suitably strong British oil company that could compete with the American companies and the Dutch controlled *Shell* on equal terms still remained. The fear that U.S. stocks were diminishing so that the U.S. could not 'be depended upon to continue her supplies at the same rate as formerly',[27] only served to emphasise Britain's precarious position.[28] It was felt that in view of the large increase in oil consumption expected in Britain after the war it was necessary to 'be prepared to substitute some other source of supply'.[29] Lord Cowdray's offer of merging his Mexican interests with those of *Burmah* and *Anglo-Persian* took on a new lease of life when he applied to the Board of Trade for permission to transfer his shares in the *Mexican Eagle Co.* to Edward I. Doheny. The Board of Trade refused permission because it was reconsidering Lord Cowdray's offer. At the same time the Admiralty proposed that *Anglo-Persian* and *Burmah* should merge.[30]

All these proposals were being considered with a view to forming a national or all-British oil company, dealing with the 'development of oilfields outside the British Isles, and in particular in British colonies, dependencies, and Allied countries'.[31] Sir Charles Greenway, Chairman of *Anglo-Persian* suggested on 4 December 1917, at the company's Annual Meeting, that *Anglo-Persian* should fill the role of the all-British oil company. Although Sir Albert Stanley (later Lord Ashfield), President of the Board of Trade, stated in the Commons on 12 December 1917, that the 'whole question is under careful consideration',[32] Lloyd George's War Cabinet later denied that it was considering the creation of such a company.[33] Sir John Cadman at a private meeting with Sir Charles Greenway also ruled out the creation of such a company as being 'contrary to national interests'.[34] Both *Shell* and *Exxon* viewed these events with irritation.[35]

Sir Marcus wrote to Sir John Cadman that it would be 'quite useless to disguise that the position of this Group might become so intolerable that they would withdraw the administration of their business from the United Kingdom'.[36] Sir Marcus was infuriated that the Group's taxes 'should be devoted by the Government to promote and encourage trade opposition to them'.[37] The proposal by the British government to acquire the British shares of *Shell* meant that British interests in the Group would exceed 50 per cent, and that the German government would now be justified in taking 'a similar course with any of the assets of Royal Dutch on which they could possibly lay their hands',[38] which would include the *Astra Romana Co.* in Romania. It was apparent then that any combination to secure the oil supplies of the British Empire that did not include *Shell* would be ineffective,[39] and that 'the only practicable course was to endeavour to bring about a combination in which British interests'[40] would secure control of *Shell*. Many of *Shell's* subsidiaries were British registered companies and most of its large fleet flew the Union Jack, but through their majority control the Dutch shareholders could at any time assign the British companies and ships to the Dutch registers.

The government considered that the essential conditions for bringing *Shell* under British control would be: first, effective British control over management and capital decisions; secondly, complete security against any transfer of this control to non-British interests; and, thirdly, that as the company or combination would be a quasi-monopoly, the interests of the consumers should be safeguarded. At this juncture British control of *Shell* was out of the question because even if consent from the shareholders could be secured, it was highly unlikely that the Dutch government would allow such a combination. However, a subsidiary company such as *Anglo-Saxon* could come into British hands by putting into the company all the Group's assets with the exception of the oilfields in the Dutch East Indies and Romania. *Burmah* would then merge with *Anglo-Saxon* and the combined British shareholding of both companies would constitute a majority British holding in the new amalgamated company. Permanent British control would be exercised by means of a voting trust, composed of five members, distributed as follows: one representative each from the British government and *Burmah*, two representatives from *Shell* and one banker. The assets of the new company so formed would be in proportion of 51 per cent British and 49 per cent Dutch, and the management of the company would be in the hands of seven directors (two from *Burmah*, three from *Shell* and two Dutch). The British government would maintain a discreet level of involvement by only retaining one vote in the voting trust. Walter Long felt that this level of involvement was not sufficient to attract *Shell* and might 'even drive them away and that means disaster'.[41]

The oil supply question was becoming more acute with the prospect of victory seeming to depend on it. Henry Bérenger, Director of the French Delegation at the Fourth Preliminary meeting of the Inter-Allied Petroleum Conference, declared that France could defend her territory only if oil supplies were guaranteed, and stated that 'for the conduct of hostilities to the mutual victory looked for with the Allies of the West, oil is recognised today to be as necessary as blood. On the battlefield, on land, on the sea or in the air, *a drop of petrol is equal to a drop of blood*'.[42] In Britain on 4 March 1918, it was decided to set up a Petroleum Imperial Policy Committee, composed of the Admiralty, Board of Trade, Foreign Office, Treasury, the Petroleum Executive and the Board of Fuel Research to 'enquire and advise as to the policy which His Majesty's Government should follow to secure supplies of oil for Naval, Military and Industrial purposes'.[43] Sir John Cadman, Director of the Petroleum Executive, in discussing the policy to be pursued stated that the present war had demonstrated the dependence of the British Empire on petroleum and its derivatives and the great difficulties it had in securing supplies. Because the U.S. supplied over 80 per cent of British requirements it was also clear that it had the power to place the UK 'in an impossible position should they desire to be unfriendly'.[44] There was consequently no time to be lost in deciding on the policy which 'will ensure to the British Empire adequate supplies of petroleum products'.[45]

On 29 May 1918 at its first meeting the Petroleum Imperial Policy Committee recommended that *Shell* should come under British control.[46] How this was to be achieved was reduced to two alternatives, either to increase British participation in *Shell* or effecting some re-arrangement of shareholdings which would change the Dutch majority in the Group to a minority. In the end a merger between *Anglo-Persian* and *Shell* was proposed so as to thwart American interests in the Middle East and Persia at the end of the war. The Dutch interests, as Sir John Cadman argued at the second meeting of the Committee, would have to be 'moulded so as to become identical'[47] with those of the British government. The question that remained was Deterding's price for coming under British control. It was felt that once certain territories and areas were handed over for *Shell* to develop Deterding would have no qualms about being British controlled. Moreover, it was to *Shell's* advantage to be 'more or less intimately connected with some great sea Power'.[48]

The negotiations, started by Sir Marcus Samuel and continued by Henri Deterding, were on the lines that *Shell* would acquire part of the government's holdings in *Anglo-Persian* or be given any additional capital that might be created *ad hoc* in that company. Deterding felt that the Dutch shareholders would more readily agree to British direction of the whole combine if they were given any additional capital that might be created in *Anglo-Persian;* for example, two million out of a total of six million shares which 'he would be prepared to pay for . . . in cash'.[49] Deterding wanted the British government's existing control of *Anglo-Persian* to remain unchanged. Oil would be supplied at competitive prices and the British Empire would have first call on any oil produced, and the Directorate of the combine would be based in London. It was conceived that this would be 'all the guarantee that was necessary or possible'.[50]

Prior to these negotiations taking place the British government in late 1917 had acquired all the 'Royal Dutch shares held by the British public'[51] for £2.7 million or £51 per share.[52] On 22 July 1918, Lord Harcourt at the Imperial War Conference reaffirmed the need to stimulate oil exploration in the British Colonies, and if necessary 'after the discovery of the oil to take Government control of the licences issued for that purpose'.[53] The conference unanimously approved the following Resolution:

> The Conference takes note of the Memorandum on the question of Petroleum,[54] and having regard to the great and growing importance of petroleum and its products for Naval, Military and industrial purposes, desires to commend the suggestions contained in the Memorandum to the serious consideration of the Governments concerned.[55]

In order to make up the deficiencies in British supplies the government started to look at alternative sources of fuel. On 19 November 1917, Walter Long appointed a committee headed by Sir Boverton Redwood to consider the employment of gas in substitution for petrol and petroleum products as a source of power, especially in motor vehicles.[56] A year later on 10 October 1918, Sir Boverton was again appointed by Long to head a committee to consider the various sources of supply of alcohol, methods of manufacture and cost, and to decide whether it was to be used in the internal combustion engine.[57] In March 1918 Churchill, Minister of Munitions, appointed a committee, headed by Lord Crew, to consider the report produced by the Petroleum Research Department on the production of fuel oil from home sources and to advise how fast it could be placed into production. The Committee found that the carbonisation of cannel coal and kindred susbstances recommended by the Petroleum Research Department was not practical but that the alternative suggestion of producing fuel oil from cannel coal and kindred substances could be developed using vertical retorts at existing gasworks. The Committee further recommended stimulating the production of Scottish shale oil, drilling for oil in the UK, the use of dehydrated tar by the employment of suitable solvents, and a great extension of the carbonisation of raw coal as a preliminary to its use for industrial and domestic purposes.[58] At the same time, after Lord Cowdray had been given assurances by the government that it would take steps to protect future British oilfields against indiscriminate drilling,[59] negotiations began with the Board of Trade for *Pearson* to put its expertise at the disposal of the government.[60] Lord Cowdray, however, wanted the government to introduce a Bill which would restrict drilling for oil.[61] As a result, Bonar Law, Chancellor of the Exchequer, on 30 July 1918, stated in the Commons that 'an endeavour has been made to settle a Bill which could be passed by general agreement',[62] with the outcome that on 1 August of the same year Sir Lansing Worthington-Evans re-introduced the Petroleum (Production) Bill. This differed from its predecessor only in that it did not contain a royalties fund[63] and that the Ministry of Munitions would grant the licence to explore for oil.[64] On 10 September 1918, *Pearson* acting as sole agents of the British government would develop 15 sites which had been approved by Sir John Cadman[65] for which they were awarded £1 million after the Bill received the Royal Assent on 21 November 1918.[66]

The Admiralty sided wholeheartedly with these views and felt that 'the gradual substitution of Oil for Coal will in the future, if the question is not immediately faced, wrest from our grasp one of the principal factors on which the maintenance of our Naval position depends'.[67] It was also prescient enough at this juncture to recommend that the government should take energetic measures to prevent the enemy from endangering the oilfields of Persia, and to push forward exploration and development of all possible oil lands in Persia and Mesopotamia by purely British interests.[68] Admiral Sir Edmond Slade's suggestion of retaining the oilbearing lands of Persia and Mesopotamia was considered by Maurice Hankey, Secretary to the War Cabinet and the Committee of Imperial Defence, to be a 'first class British war aim.'[69] In the opinion of the Army's General Staff, there was no military advantage to be gained by pushing forward to Mesopotamia. However, Hankey advised Lloyd-George, the Prime Minister, that it would be to the country's advantage to secure before the end of the war the 'valuable oil wells of Mesopotamia'.[70] The future military campaign would take this into account:

> As regards the future campaign, it would appear desirable that before we came to discuss peace, we should obtain possession of all the oil-bearing regions in Mesopotamia and Southern Persia, wherever they may be. By this time you will have read the Chief of Staff's paper on the Future Campaign, and you will have seen that both Palestine and Mesopotamia have rather become 'dead ends' because of the new line of penetration to the East which the enemy has opened up through the Caucasus towards the Caspian and from thence through Turkestan or Persia towards India. The acquisition of further oil-bearing country, however, might make it worth while for us to push on in Mesopotamia; not withstanding its comparatively minor importance from a purely strategical point of view.[71]

The question was considered by the Imperial War Cabinet and the Prime Minister's Committee which accepted the proposal.

This resolution was to have an important bearing on the negotiations going on between *Shell* and the British government. The Admiralty did not regard them as a wholly satisfactory answer as they felt that the Empire would still not be guaranteed its oil supplies. It argued that the new fields of Latin America would be 'swallowed up by the American Standard octopus, though the Royal Dutch in Venezuela and the Eagle Oil Company in Mexico will probably put up a good fight'.[72] Moreover, it was assumed that the exportable oil surplus from Central and South American oilfields would not be large enough for some years to come and 'it is only reasonable to expect that it will be absorbed by neighbouring markets in Central and South America and in the United States, and will not be available to any great extent for our requirements'.[73] The oilfields of North Borneo and Egypt were no longer in British hands and only Persia and Mesopotamia offered any hope. There is little doubt that the Admiralty reached this conclusion because of its close connection and favourable contract with *Anglo-Persian*. As a result the oil company sought to exercise its considerable influence to convince the British government to back the further commercial development of *Anglo-Persian*, while at the same time backing 'every single respectable oilfield now on British soil and to eliminate as far as possible any connections with American, Dutch or any other but British firms'.[74]

Towards the end of 1918 negotiations between *Shell* and the British government appeared to be drawing to an end.[75] In October, a proposal was made by Lord Inchcape and Sir Harry MacGowan that *Shell* should purchase half the government's shares in *Anglo-Persian* at par value. The government would retain a shilling per ton royalty on production and in return *Shell* would come under British control except for companies operating in Dutch possessions. It appeared, however, as Sir Harry MacGowan commented, that Deterding did not attach 'any importance or value whatsoever to the Royal Dutch being a British controlled company'[76] but rather wanted to acquire greater control of *Anglo-Persian*. Deterding later suggested that *Shell* should purchase a further number of shares from the government (at a mutually agreeable price) to bring *Shell*'s stake in *Anglo-Persian* to 49 per cent.[77] This offer was unacceptable to the British government because it did not want to relinquish control of *Anglo-Persian*.

With the collapse of Turkey and Mesopotamia Sir Marcus Samuel and Henri Deterding worked out a new proposal in which *TPC* should be restructured with two-thirds of its equity going to *Anglo-Persian* and the rest to *Shell*. In order to achieve this the *Deutsche Bank* shares in *TPC* would have to be acquired, and Deterding suggested that

> the present is the psychological time to get hold of the Deutsche Bank interests in the Turkish Petroleum Company which at present is . . . in the hands of the Public Trustee. If

that were done it would remove any possibility of claims being made by interested Allies at the forthcoming Peace Conference and would enable some kind of deal to be made for the fusion of Anglo-Persian and Shell interests in that particular company.[78]

The Petroleum Imperial Policy Committee took up this suggestion and recommended that the Foreign Office purchase the *Deutsche Bank* shares in *TPC* on behalf of the British government at the conclusion of the Armistice with Germany on 11 November 1918.

The situation was more complex than Deterding readily admitted. First of all, the validity of certain agreements entered into by *TPC* in 1912 and 1914 had to be resolved. Moreover, the desire of the French to secure their own sources of oil, and the wish by the U.S. to develop overseas sources of oil further complicated matters. Deterding tried to diminish the oil scaremongering by the Americans by declaring during the negotiations that 'the figures of the prospective rapid failure of North American oilfields were false, and had been "faked" for British consumption by the Standard Oil Company'.[79] The British government was aware that Deterding was trying to secure a favourable position in *Anglo-Persian* while he granted the British 'some "camouflage" control to keep us quiet',[80] but there was little that could be done to remedy this situation.

The war had demonstrated that oil was a more efficient fuel than coal. As if to emphasise this point Lord Harcourt in November 1918 at the closure of the Inter-Allied Petroleum Council made his now memorable statement that the 'Allies floated to victory on a wave of oil'.[81] The Navy was already 95 per cent powered by fuel oil and the Merchant Navy would soon follow suit, while manufacturing and heavy industry too would soon change over.[82] It was, therefore, no exaggeration when Sir John Cadman wrote that 'our existence as an Empire is very largely dependent upon our ability to maintain control of bunker fuel'.[83] Various official reports in 1919 drew attention to this crisis. Sir Wilfred Stokes' Sub-Committee of the Standing Committee on the Investigation of Prices, appointed to investigate costs, prices and profits in the oil industry, concluded that motor fuel was rapidly becoming important in all industries and that all aspects of the industry were controlled by *Exxon* and *Shell*. Sir Wilfred recommended that as a short-term measure the Board of Trade under the Profiteering Act should set maximum wholesale and retail prices for petrol in the UK and that the whole question of production, price and distribution should be put to the

Economic Section of the League of Nations for governments to settle a fair international price for oil.[84] Later that year Sir Boverton Redwood, Chairman of the Committee appointed to consider gas as a substitute for oil concluded that gas traction merited adoption in the 'propulsion of motor omnibuses and other commercial types of road motor vehicles, including tramcars'.[85]

Up to now Deterding had held the upper hand in the rounds of negotiations, but the deteriorating Dutch political situation forced him to become 'more and more British'.[86] Nevertheless, Deterding argued that the value of the shares in *TPC* was questionable as the title deeds would undoubtedly be disputed later on, and this would mean that his Dutch shareholders were being offered something which was not actually transferable. He did not think therefore that 10 per cent over *Shell's* 25 per cent interest in *TPC* would be accepted by his shareholders: only the whole *Deutsche Bank* share issue would be acceptable. He argued that:

> Anything less was not very encouraging to the Dutch. The point of view of the Dutch was that whereas the British Government wanted payment for their becoming British, other governments asked no conditions in return for the assistance of the group. The British offers did not compare at all favourably with other offers, and if he was looking only to the Companies' interests and not to British interests, he would tell them to listen no further.[87]

The Foreign Office later agreed to hand over 20–24 per cent of the *Deutsche Bank* shares to *Shell*. *TPC's* shares would then be distributed as follows: *Anglo-Persian* 49 per cent, *Shell* 49 per cent, and the British government 2 per cent. More important, however, was that *Shell* would manage the concession, something which *Anglo-Persian* opposed.[88] Britain's hand would be reinforced if at the final Peace Conference Mesopotamia were retained under a trusteeship, suzerainty or any other form. At the Conference, Britain proposed to remain on the basis of a 'Power controlling an individual Arab State.'[89] The justification for this to the other Powers attending the Conference was that it was 'primarily in the interests of the inhabitants.'[90] However, one of the first acts of the controlling Power would be to recognise the 1912 Agreement and reconstitute *TPC* to favour British interests, thus depriving the 'new State of a free hand in disposing of one of its most valuable assets'.[91]

On 3 January 1919, a provisional agreement was arrived at

between the British government and *Shell*. In return for participation in the exploitation of certain oilfields in Mesopotamia, in which the British government would control the voting power and whose management would be permanently British, *Shell* agreed to the creation of a special class of shares with special voting power to be exercised by a nominee of the British government (the Bank of England was the prospective nominee) in certain specified cases. The company also agreed that 75 per cent of the Directors of the *Shell* company, and a majority of the Directors of certain companies to be controlled by *Shell*, should be British born subjects. *Shell* undertook not to change its British directorate without the British government's approval. Furthermore, certain companies under Dutch control would now come under British control. Subject to certain exceptions, fresh acquisitions by *Shell* would be through these British controlled companies, and in order to accomplish this the Articles of Association would be changed. No interference with the commercial or financial policy or business management of the companies concerned was contemplated by the British government.[92] Under the agreement the Dutch shareholders were liable to pay British income tax. The only outstanding point which remained was the incidence of British income tax to be levied on the Dutch shareholders. Deterding realised that it would be impossible to legislate a remission of income tax for the Dutch shareholders, and felt that the only feasible way of arriving at a solution was some sort of internal arrangement between his companies. He therefore suggested that as the majority of the Group's profits was made by the Dutch producing companies, the 'Asiatic Petroleum Company should be allowed to charge a largely reduced fee for its management, and therefore increase the Dutch proportion of the profits so as to equalise matters'.[93]

The Petroleum Imperial Policy Committee, echoing the views expressed by the India Office, were worried that the exclusion of all interests except those of *Anglo-Persian* and *Shell* in the development of the Iraqi oilfields would present serious difficulties with the new government of that country. The Committee decided that this was the moment to persuade Deterding to agree to set aside 30 per cent of the shares in the new *TPC* for other interests. Deterding agreed because at the time he was having a serious dispute with the Romanian government as the latter refused to recognise the property rights of the *Astra Romana Oil Company*, Shell's Romanian subsidiary. Deterding was anxious to procure British control over the company so that the Foreign Office might exercise diplomatic pressure to have the company recognised.[94]

On 3 February, a formal provisional agreement was reached which was initialled by Lord Harcourt and Deterding on 6 March.[95] The agreement was for *Anglo-Persian* and the *Anglo-Saxon* (or any other British registered *Shell* subsidiary), to be admitted to equal participation in the oilfields held by *TPC* in Mesopotamia. The *TPC* would be reconstituted with the shareholdings divided as follows: *Anglo-Persian* 34 per cent, *Shell* 34 per cent, British government 2 per cent, others 30 per cent. The British government's shares would have special majority voting powers, exercised through a voting trust. The combined votes of *Anglo-Persian, Shell* and the British government would be placed in this Trust, and the British government would instruct the 70 per cent share block in *TPC* as to how they should vote. *Shell Transport and Trading, Anglo-Saxon* and *Asiatic* would remain British registered companies and the *Royal Dutch Co.* would cede to *Shell Transport & Trading* control of *Anglo-Saxon* and *Asiatic* as well as all subsidiaries controlled by these companies, *inter alia*, the *Colon Development Co.*, the *Venezuelan Oil Concessions Ltd.*, and the *Caribbean Petroleum Co.* in Venezuela. The *Shell* Board would be reconstituted with 75 per cent of their Directors to be British born subjects. *Royal Dutch* would retain half a million shares in *Shell* and appoint two Directors to the *Shell* Board. This provision was subject to the waiving of British income tax on Dutch shareholders. No sale of assets of any of the companies, and no change in the Articles of Association could be made without the consent of the Governor of the Bank of England, the prospective nominee for the British government. No change in *Shell's* directorate could be undertaken without the consent of the Bank. Any new interests acquired in production, transport, and marketing other than interests which might have holdings under French control and interests in territories under Dutch, American or Panamanian rule, would be acquired by *Anglo-Saxon*, unless the Bank and *Shell* consented to such interests being acquired by another subsidiary. The share capital of *Astra Romana Oil Co.* would be re-distributed so as to give equal shares to French, Dutch and British holders, whilst the Executive Board would meet in London. The Treasury later agreed that *Royal Dutch* shareholders resident outside Britain should be exempted from British income tax.[96]

The Foreign Office, Board of Trade, India Office and especially the Admiralty were in favour of the agreement because it ensured 'that a large proportion of the immense potential output of the

Royal Dutch Shell Combine will be handled by companies which are definitely and permanently British'.[97] Although the War Cabinet approved the agreement on 5 May 1919,[98] it was destined to be still-born. Deterding, a few months after initialling it with Lord Harcourt, reached the conclusion that as it stood it would carry grave dangers and risks for *Shell* owing to the 'unavoidable publicity'[99] which their competitors would propagate against the company and which would endanger its position against 'the Governments of several countries in which we are working concessions and/or in which we shall want either to obtain new concessions or to extend our existing business'.[100] He had the U.S. and Venezuela in mind. Under Venezuelan law no company, wholly or partially controlled by a foreign government, could operate in the country. In the U.S. a vicious anti-British campaign, fuelled by rumours that *Shell* was controlled by the British government, was thwarting the company's chances of obtaining new oilbearing lands. Deterding therefore set in motion a new proposal. Since *Royal Dutch* and *Shell* together held the majority of *Asiatic* and *Anglo-Saxon*'s shares, Deterding realized that owing to the Dutch majority shareholding in *Shell*, the Board of Directors to be elected could be either Dutch or at least anti-British. He therefore proposed that:

> If, however, arrangements were made by which the General Meeting of shareholders were bound to appoint members of the Board in such a way that the majority should be British; and if, in addition, the choice of the British members should be subjected to the approval of the British Government through a nominee if desired, then in our opinion the same safeguards as to British control of the business would be achieved.[101]

Moreover, shareholders could exercise control over the business only through the nomination of Directors and management. If sufficient safeguards were provided for the Board and consequently the whole management to be in the hands of British born British subjects (a safeguard approved by the government), it did not matter whether *Royal Dutch* or *Shell* had a majority at the General Meeting of shareholders. The more so when under the initialled agreement the veto right was held by the British nominee. It was felt that the

> General Meeting of shareholders could not make a decision without the concurrence of the minority (by inserting in the

Articles of Association that such decisions would require a three-fourths majority) whilst the minority i.e. 'Shell' Company would then enter into an understanding with the British Government by which the 'Shell' would not consent to any such decision without the approval of the British nominee.[102]

The Articles of Association of the *Anglo-Saxon* would be changed to stipulate that the majority of the Directors must be British born subjects, and that certain decisions could only be taken with a three-fourths majority. *Shell* felt that these stipulations would not arouse any criticism from the public or foreign countries as they would seem reasonable where British companies were concerned, and where the 'minority have such a large interest in the whole business'.[103]

The above proposals and the position of the Dutch shareholders with respect to British income tax were incorporated within a new draft which was sent to the Petroleum Executive for study. The last two articles included in the draft appeared to forestall efforts to reach an agreement: article 19 stated that the agreement would have to be ratified by the General Meeting of *Shell* shareholders, and once ratified, under article 20, *Shell* was bound to carry it into effect when the British government had secured *Shell*'s participation in the Mesopotamian oilfields. This last proposal was quite 'unreasonable as the Shell shareholders might refuse to pass the necessary resolutions to give effect to the Agreement'[104] after *Shell* had secured participation in Mesopotamia. In order to strengthen the position of the British government, Harold Brown, a legal expert, recommended that the draft should be amended to contain a provision in the Articles of Association of *Shell, Anglo-Saxon* and *Asiatic* and other subsidiary companies precluding the making of long-term agreements for the disposal of their produce without the approval of an Extraordinary Resolution of shareholders. Moreover, the whole of *Shell*'s holdings in *Anglo-Saxon* and *Asiatic* might be put in a voting trust framed to secure and resolve all questions dealing with appointments or removing Directors, disposing of capital assets, long-term arrangements for disposal of production, and liquidation which might prejudicially affect British control. The votes would be cast jointly representing *Shell*'s interest, but on any other question each party would have the right to decide individually. The question of appointing the Directors of *Anglo-Saxon* was discussed at a meeting held on 16 December 1919 between Dr. Stuart, Mr. Colijn (representing *Shell*) and the

1. William Knox D'Arcy – c.1907. A wealthy Englishman who made his fortune from gold mining in Australia, and obtained a 60-year-concession to search for oil in Persia.

2. Persia – H.I.M. Nasir al-Din Shah. He granted several concessions
which included oil rights before his successor, Muzaffar al-Din, signed
the D'Arcy Concession on 28th May 1901.

3. Mr C. Greenway at Sar-i-Pul, on his way from Chiah Surkh to Tehran – 1911.

4. Sir Robert Waley Cohen.

5. Sir Henri Deterding
(1865–1939).

6. Marcus Samuel,
First Viscount
Bearsted (1853–1927),
joint founder of the
Shell Transport and
Trading Company.

7. Sir John Cadman
seen before leaving
Tehran airport at the
conclusion of
Concession
negotiations – 1933.

8. Mr. Reynolds, Mr. Willans and Mr. Crush having lunch, Persia c.1910. D'Arcy employed George Reynolds, (far left), to conduct drilling operations. Funding was assumed by the Burmah Oil Company in 1905, and Burmah founded Anglo-Persian in 1909.

9. Early geological survey party in Persia.

10. 'Colonel' Edwin L. Drake talking with Peter Wilson (left), a Titusville, Pennsylvania, druggist. On the extreme right in the background is 'Uncle Bill' Smith, Drake's head driller. This is where the modern oil industry started.

11. D'Arcy's first well – Chiah Surkh. Chiah Surkh was the scene of the first drilling in 1902. Oil was eventually struck in commercial quantities at Masjid-i-Sulaiman on 26th May, 1908.

12. No. 1 discovery well, Masjid-i-Sulaiman, Persia. A 'gusher' with wooden derrick typical of the period when precautions to control the flow of oil were not always adequate.

Petroleum Executive. Stuart and Colijn stated that if *Shell* were allowed to appoint the Directors it would lend weight to the suggestion that the Group was under British government control. It was eventually agreed in principle that two Directors of these companies should be nominated by *Royal Dutch,* subject to approval by *Shell,* and that of the remaining Directors, one half would be nominated by *Shell* and the other half by *Royal Dutch.* At all times the majority of the Board would be made up of British subjects by birth.[105]

However, the British government was still not satisfied with the proposed agreement. Article 1 stated that the British government would support the reconstituted British *Shell* to obtain oil concessions and trade facilities in all territories under British rule or influence. On 26 March 1920, it was agreed that the company would be given the same measure of diplomatic support as other British companies, and the British government also acquiesced on the following points: (a) it would endeavour to obtain for *TPC* the sole right to exploit and work all oilfields in Mesopotamia which were included in *TPC's* original concession; (b) to endeavour to obtain recognition of the *TPC* rights which may not come under British control; and, (c) to give diplomatic support to the company in obtaining oil concessions and trading facilities in other parts of Asia formerly under Turkish rule.[106] Despite agreement on those points, there were serious doubts as to whether any other party aside from *Shell* and the British government would recognise the validity of the agreement as it stood because it purported to 'bind various parties who are not party to the Agreement to do various things'.[107] Harold Brown believed that the government would have great difficulty in making this agreement stick against *Shell* especially, for example, as specific points of the contract were unenforceable under English law. Thus:

> the whole Agreement will depend for its validity upon getting the Board of Shell constituted of British born British subjects approved by the British nominee, and it will then be necessary to rely upon these Directors to see that the Agreement is lived up to. Unless the Government is prepared to do this it seems to me that they should recognise that the Agreement will not secure the object they have in view.[108]

A completely watertight agreement could be achieved if the *Anglo-Saxon* and *Asiatic* shares, including those held by *Royal Dutch,* were placed in a voting trust under British government control.

FRENCH OIL POLICY

British strategy, so far, was based on the principle of allowing British interests to develop the Mesopotamian oilfields. Britain's claim was based on the shady concession title held by *TPC*. Owing to treaty obligations with the French over spheres of influence in the Middle East, consent would have to be reached with them before *TPC* could proceed with development work. In oil matters the French were also seeking to acquire independent oil supplies but were in a weaker position compared to the British.

Under the Sykes–Picot Agreement of 1916 between France and Britain, the Middle East was divided into spheres of influence with the Mosul Vilayet coming under the French. Towards the end of the war France allowed Mosul to come under British influence, while Britain recognised France's right over Syria, as well as a stake in the oil produced from Mesopotamia. France herself did not possess the expertise to develop the oilfields, but she was not unduly concerned at the time about this as it was felt that any company operating in Mesopotamia (as well as *Anglo-Persian)* would need to build a pipeline to the Mediterranean for oil to reach market. Agreement with France over the construction of the pipeline would be essential since Syria was to become a French Mandate. Moreover, *Shell* would only consider becoming British once an agreement had been concluded with the French over the Mosul Vilayet.

Informal talks took place in Paris during December 1918 between Senator Bérenger, Commissary-General for Petroleum in France, and Sir John Cadman, Director of the Petroleum Executive, on behalf of Walter Long, Minister in Charge of Petroleum Affairs. It was suggested that French interests should acquire either directly or through *Shell* the *Deutsche Bank* shares in *TPC*.[109] The position of the French government was clearly stated in a letter sent on 6 January 1919, to the British Foreign Office advising that France wanted to secure her oil supplies in a manner similar to the way in which the *Anglo-Persian* deal had been carried out. The intention was not to form an oil monopoly in a definite locality but rather to acquire shares in a company which was developing petroleum sources in different countries.

On 15 January 1919, a meeting of Under-Secretaries, presided over by Walter Long, discussed the French proposals. It became clear that the French government was negotiating not only with the British but also with the U.S. Owing to Britain's overdependence on American oil, the prospect of an American company entering the Middle East was not viewed with equanimity. It was therefore decided to come to terms with the French as soon as possible before American assistance had grown any further.[110] French participation in the *Anglo-Persian* was ruled out altogether but on 1 February 1919, at a meeting in Paris chaired by Sir Hubert Llewelyn-Smith of the Board of Trade, it was decided that France would either participate to the extent of 25 per cent of the equity of the company formed to exploit the Mesopotamian oilfields or be entitled to 25 per cent of the oil produced should the British government decide to form a company. On 6 March 1919[111] Walter Long and Bérenger initialled the first draft of the Memorandum dealing with the oil policy to be pursued by France and the UK in Romania, Mesopotamia, North Africa and British colonies. It was later sent to the Foreign Office, India Office, Board of Trade, Colonial Office and the Admiralty for their approval. Three weeks later on 29 April 1919, at a meeting chaired by Lord Curzon, the Foreign Secretary, at which all the above departments were represented, it was resolved to recommend that the Cabinet approve the Anglo-French Petroleum Agreement in view of the *Shell* agreement with the government. However, on 20 August, the War Cabinet deemed the proposal unacceptable. The Petroleum Department were surprised at this decision and felt that the Cabinet had not given sufficient thought to the consequences which such a rejection entailed. It therefore pressed the Cabinet to reconsider its decision, and a Memorandum detailing the advantages to Britain of entering into such an agreement was prepared in which it was pointed out that accord had to be reached with the French before *Shell* could come under British control. There was a strong possibility that if the Agreement were not concluded France would seek U.S. help;[112] moreover, the need to find new oil sources was now more pressing because of the general belief that American oil stocks were diminishing. The proposed agreement with the French would enable

> pipelines to be constructed from the Persian and Mesopotamian oilfields to the Mediterranean coast, and provides for the construction of refineries, and everything necessary for the establishment of large oil bunkering stations on the Eastern Mediterranean.[113]

and

> The overwhelming extent of our dependence on the United

States of America for fuel oil during the War, and the rapidly growing use of oil in the mercantile marines makes it important that every endeavour should be made to open up sources of supply under British control, and if the French Agreement, with the facilities it gives for pipelines, refineries, depots, loading wharves etc. is not satisfied, it is probable that such facilities with French spheres of influence would be secured by American interest and the full advantages of our hold on the Persian Oil Fields would not be obtained, no direct outlet to the Mediterranean would apparently be obtainable.[114]

The Agreement would therefore help to alleviate Britain's over-dependence on U.S. oil supplies. Lloyd George's new Coalition Cabinet reconsidered the proposals and authorised French interests a 25 per cent share in the area's oil. On 23 January 1920, it also concluded that, as a matter of principle, the revenues arising from the exploitation of the oilfields of Mesopotamia would accrue to the Arab state. This decision was reached in order to involve the new Kingdom of Iraq (which became a British Mandate at the Treaty of Sèvres signed in San Remo in 1920)[115] in the exploitation of its own natural resources. At the same San Remo Conference on 24 April 1920, the Anglo-French Petroleum Agreement was signed which, in addition to pledging mutual support in any negotiations over Romanian oilfields and in obtaining oil concessions in the U.S.S.R., ensured that the Mosul and Iraqi oilfields would remain under British jurisdiction. If the oilfields were to be developed by a government agency then the British government undertook to provide the French government or its nominee with a quarter of the net crude output of the region. If on the other hand a private company was to be formed to exploit the oilfields then the French would receive a 25 per cent share in such a company, with native interests receiving 20 per cent maximum participation in the company's share capital. The French would contribute one-half of the first 10 per cent of such native participation and the additional 10 per cent would be provided by each participant in proportion to its holding.[116] France agreed to the construction of two separate pipelines and railways through French Mandated Syria and would be entitled to purchase 25 per cent of *Anglo-Persian*'s oil transorted to the Mediterranean.

The San Remo Agreement came as a bombshell to the U.S. government because American interests had been barred from participating in Mesopotamia.[117] As a result, the U.S. challenged the legality of the agreement, arguing that before the war *TPC* did not hold a concession for exploiting oil. But the U.S., as Shwadran explains, was in a peculiar situation in Mesopotamia as

> It had never declared war on Turkey; it had not participated in the negotiations on the territorial disposition of the Ottaman Empire, and therefore took no part in the peace negotiations with Turkey. Nevertheless, it felt very strongly that since it had contributed materially to the Allied victory over Turkey, it was entitled to the fruits of victory, and that Americans should be given an equal opportunity with the other victors of developing the economic possibilities of the mandated territories.[118]

American protests served to delay recognition of the Agreement by the League of Nations. However, the French were pleased about U.S. protests, and on 14 December, Secretary of State Colby was so advised.[119] Their hope was to cajole the U.S. government into insisting on modifications which would be to France's advantage. In order to allay U.S. fears Sir Auckland Geddes, British Ambassador at Washington,[120] informed Secretary of State Colby that the agreement had been drawn up merely to secure French participation (at ordinary commercial rates) in the area's oil production. Philip Kerr, a Private Secretary to Lloyd George, notified Lord Curzon on behalf of the Prime Minister that the agreement was not designed to 'divide the oil between Great Britain and France'.[121] Britain, as the Power responsible for developing the Mesopotamian oilfields, had entered into an agreement to guarantee oil supplies to France, and this did not preclude the participation of other nations. The same applied to another area where the U.S. government conceived its citizens had been barred from entering: Persia.

PERSIA

With the collapse of the Russian government in 1917, and the defeat of the Turks during the First World War, Britain found herself as the only major Power in Persia, and was determined to establish herself militarily. As part of this process the British persuaded the Shah to dismiss Premier Ala-es-Sultana Dowleh and appoint Vossugh-ed-Dowleh in his place.[122] An Anglo-Persian Agreement was then concluded in 1919 which allowed the British to supply military advisers, arms and ammunition to Persia at the government's own expense.[123] Opposition to this agreement came not only from Persia itself but also from France and the U.S. Both nations felt that the end result of the agreement would be the creation of a new British Protectorate, and the U.S. viewed this as another example in which American enterprise would be left out in the cold. Britain did not recognise the Persians as participants in the Paris Peace Conference, but the U.S. and France both gave advice to them. Persian opposition resulted in the formation in 1920 of a new government which refused to act on the agreement until it had been ratified by the Majlis (the Persian Parliament) and refused therefore to accept the proposed British £2 million loan. Reza Khan, then Minister of War, succeeded in getting the 1919 Anglo-Persian Agreement rescinded, while at the same time the Russo-Persian Agreement was ratified by the Majlis. Reza Khan then turned his attention to the revenues accruing from *Anglo-Persian*'s activities. Under article 10 of the D'Arcy concession the Persian government was entitled to 16 per cent of the annual profits. However, during the First World War the oil was sold at discount rates to the Navy and the question of profit sharing from *Anglo-Persian*'s subsidiaries outside Persia had prevented the Persian government from obtaining its full share of the profits. Reza Khan engaged the services of W. McClintock, a London chartered accountant, to look into *Anglo-Persian*'s concession. Since the end of the war *Anglo-Persian* had pursued a different policy from that followed in its pre-war days. It strove to integrate vertically and acquired a number of companies and concessions in other parts of the world. At the same time its subsidiary *D'Arcy Exploration* was 'actively engaged in searching countries on the quiet for oil'.[124] McClintock took this into account when looking into the company's affairs and concluded that the Persian government was not receiving its proper share of the profits.[125] On 22 December 1920, the Armitage-Smith Agreement was signed between the Persian government and *Anglo-Persian* in which the Persian government would receive 16 per cent of the annual profits of *Anglo-Persian* companies other than those formed to transport oil by sea. The agreement rendered the Persian government's revenues more sensitive to a decline in *Anglo-Persian*'s income than before.

NEGOTIATIONS FOR BRITISH CONTROL OF SHELL CONTINUE

A second report on motor fuel prepared by Sir Wilfred Stokes and approved by the Standing Committee on the Investigation of Prices on 27 November 1920, painted an alarming picture when it concluded that as a result of a lack of development of new oil sources in the U.S., Britain could no longer rely on this source for its estimated needs of 100 million barrels per annum. Other traditional sources such as Romania and Russia could not complement this shortfall because they had also increased their internal consumption. In addition home efforts had been disappointing. Both the *Oilfields of England Ltd.*[126] and Reginald Gilbey[127] who had acquired licences to drill and exploit oil and gas in 1919 and 1920 respectively were unsuccessful. *Pearson,* however, in May 1919 found oil at Hardstoft, Derbyshire, at a depth of 3075 feet,[128] which flowed at a rate of 260 gallons per day,[129] but by the end of 1920 had spent nearly half the available £1 million[130] which Parliament had provided without finding any commercial oilfields. By the early 1920's there had been 11 borings, seven in Derbyshire, two in North Staffordshire and two in Scotland. There was also the problem that before production could be stepped up on a commercial basis the necessary legislation had to be enacted by Parliament.[131] Shale oil, which had been mined in Scotland since 1878 and had produced 3 million tons of shale per annum, was not seen as an alternative because of its high price and availability of alternative sources. The real problem, as Sir Wilfred Stokes saw it, was that 'powerful combinations, whose financial resources are enormous . . . own the principal sources of supply, and their possession of the bulk of the distributive machinery renders effective competition impossible'.[132] It was recognised that competition would not check the power of these companies; moreover, government intervention in setting prices for products on the basis of costs of production would only divert supplies to other markets. Nevertheless, Sir Wilfred felt that the government could 'control freight and fix fair distribution and other charges in this country'.[133] The long-term solution was for the League of Nations to take the decision to encourage and stimulate the production of substitutes, especially coal, which would not be controlled by monopolies.[134]

Nevertheless, the British government persisted with its desire to secure British control over *Shell.* Although there had been disagreement as to the terms of the proposed merger of *Shell* with *Anglo-Persian,* there was little doubt among government officials that the merger was to Britain's advantage. The increase of seven pence per gallon in the price of petrol effected in March 1920 and the ensuing correspondence over this between the Board of Trade on the one hand and *Shell* and *Anglo-American* on the other, had, according to Sir Robert Horne (newly appointed President of the

Board of Trade), 'brought out in a most striking way our
dependence on these two international groups of capitalists for what
has come to be one of the essentials of our civilization',[135] and he
consequently felt 'strongly that we ought not to allow the
opportunity to bring one of the groups under British control to be
missed'.[136] He also felt that the British government would be unable
to develop the Mesopotamian oilfields on its own because it lacked
the 'necessary organization for so vast a business as the successful
commercial exploitation of a large oilfield and the marketing of its
products, and I doubt if it can create it except at great and
prohibitive expense'.[137] *Shell* on the other hand did possess the
necessary expertise and capital and wanted some form of co-
operation with *Anglo-Persian* to forestall American competition.
However, negotiations leading to such an arrangement were
delayed by over a year.

In June 1921 *Anglo-American* announced that they would raise
additional capital of £5 million in London to finance the extension
of their retail outlets. *Shell* felt that this move was designed
primarily to keep *Anglo-Persian* out of the retail markets. But *Shell*,
if it was to maintain its market position, would have to match
Exxon's investment. At the time *Anglo-Persian* was also looking for
a way to extend its retail network, with the result that on 6 July
1921, *Shell-Mex Marketing Co.* proposed to *Anglo-Persian* the
setting up of a joint distribution network. This would mean that
each company would have to invest only £2.5 million in order to
match *Exxon*'s expansion plans.[138] A meeting at the Board of Trade
took place to discuss the joint venture. Sir Philip Lloyd-Greame,[139]
President of the Board of Trade, felt that though the deal was
'perfectly sound from a business standpoint'[140] it would be 'bitterly
attacked in Parliament by the public, who had a prejudice against oil
companies and combinations and did not recognise the big work
which the oil companies had done'.[141] It became apparent at the
meeting, however, that the companies 'had a wider scheme of
combination in view',[142] which was for *Shell, Anglo-Persian* and
Burmah to merge. The advantage of such a merger for the British
government was that a British majority shareholding would acquire
the commercial and financial control of *Shell*, thus ensuring greater
security of supplies and also provide a strong British company
which could compete against *Exxon* successfully. Because of a 50
per cent reduction in capital expenditure and running costs, prices,
too, would come down, benefiting the consumer. The advantages
for *Burmah* were that it broadened 'the whole basis of their business

and of their security in that it gives them an interest in world wide fields and markets in place of their dependence on the oil fields of Burma, India and Persia'.[143] The way proposed to secure British control of *Shell* was through a combination of the British shareholdings of the three companies as shown in Table I.

TABLE I

Proposed merger between Shell, Anglo Persian and Burmah

Company	Ordinary £1 share price	Number of shares (millions)	Cost (£million)	Percentage distri- bution
Burmah	£6 5/8	5.2	34.1	10.7
Shell	£5 10/-	19.3	106.3	33.2
HMG shares in *Anglo-Persian*	£4	5.0	20.0	6.3
Royal Dutch (60% of Group)	–	–	159.4	49.8
TOTAL	–	–	319.7	100.0

Source: POWE 33/92 J.C. Clarke to W.St.D. Jenkins, 11.10.21

As we can see British shareholding in the merged group would be 50.2 per cent while Dutch shares would represent 49.8 per cent of the total value of the proposed merger. British control would be secured by special provision in the Articles of Association for a voting trust or whatever formula legal advisers recommended.

In October a further meeting was held at the Board of Trade to discuss the proposal. W. St.D. Jenkins of the Admiralty viewed with trepidation the extinction of competition in fuel oil in the East, but was nevertheless willing to sanction the proposal once its bunker oil was guaranteed at a low price.[144] Sir Hubert Llewellyn-Smith felt that the establishment of an 'effective British control over the Royal Dutch-Shell group was worth paying a substantial price for',[145] but that the proposal did not give sufficient details as to how effective control of *Shell* would be achieved, and warned that 'pressure of events might at any time lead it [*Shell*] to accept American control and to form a world group with the Standard Oil Company'.[146] The proposed amalgamation would therefore put to an end any notions which the Americans had of acquiring *Shell*,[147] and it was agreed that the merger would be in the very best of British interests. The meeting proposed to ascertain from *Burmah* (which had formally submitted the proposal) more details on how to gain effective control of *Shell*. Robert Watson, *Burmah*'s representative, answered that there were two ways open to secure this end: one was for *Burmah* and *Shell* to merge, and the other was to merge *Burmah* and the British governement's shares in *Anglo-Persian*, with *Shell*. In either case *Shell* would retain its identity as a holding company and would acquire an interest in all the operating subsidiary companies. Equally, either alternative would deplete the 100 per cent British element in *Anglo-Persian* and *Burmah*.

The absolute guarantee demanded by the Admiralty that there should be an obligation to supply fuel oil at low prices could not be undertaken because of the limitations (political instability and logistics) of working in foreign countries. Watson finally warned that:

> No voting trusts, no formal agreements, no anything can entirely overcome these physical inherent obstacles to complete British control. Either grouping, however, increases the good will which ordinarily would naturally react in favour of, and to the advantage of, British interests.[148]

The only attractive feature of the scheme was the possibility of some sort of British control over *Shell*. The Petroleum Department rejected Watson's second alternative as 'being entirely out of the question'[149] because the government would lose its shares in *Anglo-Persian*.[150] The first alternative was also deficient because it would give *Shell* control of *Burmah*'s equity in *Anglo-Persian* and a 'disproportionate share in the direction of its commercial policy'.[151] The merger would involve either the British government's participation in an international oil trust or the sale of its shares in *Anglo-Persian*, both alternatives being politically unacceptable. The Petroleum Department concluded that the government would be unable to answer criticism 'which would certainly be evoked by an action so diametrically opposed to that taken in 1914, and by the formation of an immensely powerful trust which could take place with Government approval and assistance'.[152] The creation of such a new company would reduce the number of large independent marketing groups in the UK from three to two, and the new company would be viewed by smaller countries with the same distrust as *Exxon*.

The Admiralty for its part did not favour the merger because its supplies and price requirements would not be met. Its contracts with *Anglo-Persian* and *Burmah* had a long time to run and offered better prospects. Moreover, after the war *Anglo-Persian* had extended its operations outside Persia. In Venezuela, for example, the company now owned, through subsidiaries, a number of concessions.[153] On 28 August 1920, the *British Equatorial Oil Co. Ltd.*[154] was registered in London with a capital of £200,000, subscribed by the *National Mining Corp. Ltd., The Mexican Corp. Ltd., The D'Arcy Exploration Co. Ltd. (Anglo-Persian's* subsidiary), and the *Scottish American Oil & Transport Co.,*[155] acquiring around 170,000 hectares in Venezuela.[156] *Anglo-Persian* also had a stake in the *North Venezuelan Petroleum Co. Ltd.,* registered in London on 10 February 1920, with a capital of £300,000,[157] and jointly owned by *Anglo-Persian, Central Mining & Investment Corp.,* and the *Trinidad Leaseholds Ltd.* to acquire the Jiménez Arraiz concession in Venezuela. In 1922 A.C. Hearn secured an option for *Anglo-Persian* to purchase 1,200 hectares of oil lands from the *Mara Exploration Oil Corp.*[158]

After two years of negotiations Deterding was annoyed that no formal agreement had been reached.[159] He decided to pursue a tactic that would create doubt in the British government as to the consequences of not accepting the agreement. Colonel Boyle, Deterding's secret agent, saw J.C. Clarke of the Petroleum Department and informed him that Deterding was the only obstacle to the sale by Dutch shareholders to *Exxon*. Once Deterding retired, Colonel Boyle stated, the Dutch 'would join forces with Standard within two or three years'.[160] Deterding, however, had misread the way the British government was moving as British oil policy at this juncture was moving towards forging closer links with *Exxon*. There was little doubt that the proposed *Shell* merger

> would be interpreted in the United States as evidence of a determination to build up a British organisation strong enough to fight and beat the Standard Oil Company. The latter has . . . been extending its interests in other American oil groups, is wealthier and more powerful than ever, and has been making its influence felt to our detriment in various countries where it is by no means scrupulous as to the weapons it employs. The announcement of the amalgamation would, therefore, probably excite angry criticism in the United States and considerable friction might result for a time; but if a disposition was shown to come to terms with the Standard in countries where they desired participation, such as North Persia, Roumania, and Mesopotamia, any active irritation would probably be only short lived.[161]

Negotiations between *Anglo-Persian* and *Exxon* to enable the latter to buy part of the former's production had started. Moreover, accord over Northern Persia was drawing to a conclusion, and according to Sir Philip Lloyd-Greame it was sound policy for *Anglo-Persian* and *Exxon* to have links because 'such an association would automatically reduce the likelihood of an alliance to our disadvantage between *Royal Dutch-Shell* and the *Standard*'.[162]

In late January 1922, Deterding was called to Holland for a meeting with the Dutch Council of Ministers. He was sure that this was 'to make peace proposals both as to export duties and as to other matters which have been creating friction between Royal Dutch-Shell and the Dutch Government'.[163] Deterding felt that once these questions had been resolved the opportunity for reaching an agreement to merge would have ended, because his Dutch colleagues agreed to the merger only as retaliation to the Dutch government's lack of interest in the company's affairs. Watson felt that by using this ploy Deterding was attempting to 'force the pace of the scheme over here'[164] because once the merger had been agreed Deterding would face his Dutch colleagues with a *fait accompli,* thereby forcing its acceptance. However, no promise of an agreement was given by the British government. Instead, Lloyd George appointed a Committee, headed by Stanley Baldwin, President of the Board of Trade, to consider the proposed merger and the policy to be adopted with regard to the British government's shares in *Anglo-Persian*.[165]

THE U.S.S.R.

Deterding sought British control over his company because, with Foreign Office backing, he might be able to arrive at a favourable agreement with the governments of Russia and Romania about the considerable assets *Shell* held in those two countries.

In 1912 *Shell* acquired 80 per cent of the Rothschild Frères holdings in Baku for shares in *Shell* worth £241,000 and for shares in *Royal Dutch* worth 3.9 million guilders. *Shell* also acquired the Rothschild distribution agencies throughout Russia as well as the subsidiaries at the Grosny and Emba (Urals) oilfields. After the war, it procured smaller Russian companies such as the Lianoroff and Mantacheff interests in the Grosny district. It also obtained the Nikopol-Mariupol pipe factory for £250,000, the largest pipe factory in Russia.[166] The British too were interested in fostering stronger trade links with the new Russian government of Lenin. In

1921, Lloyd George signed the Anglo-Russian Trade Agreement and *Anglo-Persian* was encouraged to purchase oil from the Baku oilfields. Despite this, Russia concluded that owing to high production costs, the lack of an adequate transport system, and a shortage of capital, outside help to develop the fields would be needed.[167] The first Russian scheme put forward was to unify all oilfields and then divide them up into shares with 25 per cent going to the Russian government, 25 per cent to the previous company owners, and 50 per cent to American, British and Dutch groups to provide working capital. At the first Pan-Russian Oil Congress this scheme was rejected.[168] A more positive effort was made in April 1922 during the Genoa Conference,[169] which became the Mecca for oil men who wished to participate in Russian oil developments.

The prize was a large one because Russia was the second largest oil producer in the world. At first *Exxon* showed little interest but changed its mind when *Shell*'s and Russia's secret proposals for an oil development agreement leaked out. *Exxon* together with the French and Belgian *Nobel* shareholders took up the intransigent position that if they were not permitted to develop the *Nobel* properties, then all other companies should refrain from dealing in Russian oil. Rumours spread that the British government was using its influence to acquire for *Shell* large concessions in Russia, which were denied by both Colonel Boyle, *Shell*'s special agent, and Krassin, Russia's representative. At the time Krassin was working on a resolution to divide oil production equally among Belgian, French, American and British interests as well as former Russian owners who would make up a consortium. *Shell* opposed this as it would give equal status to its main rival *Exxon*. On 2 May an agreement was reached between Britain and Russia whereby *Shell* would develop its oil properties held before the Revolution. A similar understanding could not be reached with *Exxon* as this agreement was only binding on properties held before nationalisation, and *Exxon* had acquired its properties in 1920. Nevertheless, the Americans together with the Belgians and French, opposed *Shell*'s scheme vigorously, with the result that the Conference closed without reaching any agreement.

The discussions continued later at The Hague Conference held on 19 June 1922. The Russians were not willing to discuss oil matters before clear assurances had been given that Russia would receive loans or credits or both. Having received these, the Russians stuck to their original plan of an all-inclusive merger of all interests, instead of the granting of many concessions. The oil companies were unwilling to co-operate with each other on this matter, and the conference ended with an informal agreement to establish an international blockade of Russian oil. This would be difficult to achieve because of the general belief that world oil stocks would not last long and because the depletion of these stocks was taking place at a much faster rate as a result of the large post-war demand for oil. As a result, in September the oil men gathered in Paris for a further conference to establish sanctions which would stop companies from purchasing Russian oil. According to Fischer the real object was to 'tame the Royal Dutch, which alone among the petroleum powers would not desist from its efforts to regain a foothold in the Caucasus'.[170] The companies formed into a *Groupement Internationale des Societés Naphtières en Russie* in order to ensure a strong blockade of Russian oil. The fact that *Exxon* and *Shell* formed part of the *Groupement* militated against its success, and Russian oil exports steadily increased over the years, 55 per cent of its products reaching Britain. It was said that only small independents bought Russian oil, but when an independent trader acquired 30,000 tons of Russian kerosene, *Shell* suspected *Exxon* of foul play. As a result *Shell* placed an order for 70,000 tons of Russian oil with *Arcos Ltd.*, the Soviet trading organization in London. This eventually led to the breaking up of the *Groupement*, and the Russian government decided to form its own company, the *Russian Oil Products Ltd.*, to market and distribute its oil overseas.

BALDWIN'S OIL COMMITTEE

Baldwin's oil Committee considered the proposed merger between `Shell, Anglo-Persian` and *Burmah*. But before any conclusions could be reached American, more specifically *Exxon's* reaction to the merger, had to be taken into account. On 13 March 1922, Sir Arthur Balfour, Lord President, stated to the Committee that *Exxon* 'had carried out very powerful propaganda and as a result had obtained the sympathy of an ill-informed State Department – their whole aim being to arouse public opinion in the United States against Great Britain on the ground that Great Britain was trying to gain a monopoly of the oil supplies of the world'.[171] Oil was therefore a 'most fruitful and easy field for propaganda'.[172]

This view was confirmed by a meeting between Sir Auckland Geddes and Alfred Bedford, president of *Exxon*, held in Washington. Sir Auckland reported to the Foreign Office that:

Bedford is at present personally most friendly, and is now publicly committed to full support of Anglo-American friendship. Speaking in favour of its full development, he said that he deplored the bad effects that any such amalgamation must have on American public opinion, and stated with an air of sadness his conviction that neither he nor I would be able to allay the storm which would break out in the press if such an amalgamation were announced, or if its negotiations were suspected. This is . . . plain warning that if the Anglo-Persian Oil Company passes under the control of Royal Dutch-Shell, or if their interests are combined, or if Bedford thinks this likely to happen, the full resources of Standard Oil Company's propaganda machine in the United States and in France, Italy and other European countries, as well as in Asia, will once again be directed against Great Britain.[173]

Sir Auckland further urged Bedford 'with whom the decision rests to weigh the prolific effects that a ruthless anti-British propaganda, run in every part of the world that *Standard Oil Company* can reach, may have upon British policy at a time like this'.[174] The Admiralty, too, voiced its disapproval of the merger as they would not be guaranteed their oil. The Committee's final report acknowledged its inherent commercial advantages, particularly in reducing capital costs. It stated that:

The Royal Dutch and Shell Companies would secure large and increasing supplies of cheap oil which would be marketed by their existing organisation; the Anglo Persian Company would avoid the necessity of creating costly facilities for distribution; while the Burmah Oil Company, at present mainly dependent on a single oil field, would spread its risks. The latter Company as a large shareholder in the Anglo Persian Oil Company, has also insisted on the commercial objections to having a large and expensive distributing organization based on production which is so far confined almost entirely to a comparatively restricted area in Persia.[175]

However, it was felt that the merger would stifle competition and some of the smaller companies would go out of business. It would also mean that companies could adjust supply and demand and fix prices to suit their requirements. No government interference in this field would be welcomed by the companies.

The Committee, furthermore, was not satisfied that British

control over oil properties in foreign territories could be exercised permanently under the amalgamation proposal because 'in the last resort effective control over oil companies in time of war can only be obtained by control over production in the producing territory'.[176] British control of the companies' policy or the election of Directors could only be secured by creating shares of a nominal amount with special voting powers that could be assigned to nominees approved by the British government. However, the 'real management of the amalgamation will inevitably pass to the man of outstanding ability and keenest brain, whatever his nationality'.[177] In the strongest terms possible the Committee recommended that the British government refuse to grant permission for *Anglo-Persian* to enter into such a merger with *Shell,* and that it should retain its equity holding in *Anglo-Persian.* This put to an end any notion of *Shell* coming under British control.

During the last two years of World War I it became obvious to the British government that its future oil policy would have as its overriding objective the procurement of independent sources of oil. This necessitated a strong British company as it appeared that neither *Burmah* nor *Anglo-Persian* possessed the 'economic independence, the areas of supply or the commercial ability to enable them to fulfil the necessary conditions'.[178] It would be too costly to set up a wholly independent British company to compete successfully against the giant oil companies. Thus, during the last stage of the war and after the end of hostilities the British government pursued a vigorous policy through the Petroleum Executive and its off-shoot, the Petroleum Imperial Policy Committee, to cause *Shell* (sixty per cent owned by Dutch interests) to become a wholly British controlled company, with the government retaining certain voting rights. Although a draft agreement was initialled by Lord Harcourt (President of the Petroleum Executive) and Henri Deterding (head of *Shell)* on 6 March 1919, no final agreement was reached and negotiations were discontinued in 1922. The policy to secure Britain's oil independence had failed; what was even worse was that the Persian and Iraqi oilfields, expected to supply the majority of British needs, were not developed sufficiently fast as a result of local political and international complications.[179] The U.S. up to 1932 still remained Britain's largest oil supplier.

NOTES

1. Hereinafter *Asiatic*.
2. Hereinafter *Anglo-American*.
3. Royal Commission on Dominions, *op. cit.*, Tables 130-2, pp.462-3.
4. *Ibid.*, Table 133, p.463.
5. *Hansard*, 1916, Vol. 80, Col. 2714, Mr. C. Watson, 31.5.16.
6. *Hansard*, 1916, Vol. 80, 28.6.16.
7. POWE 33/1 Board of Trade, Petrol Control Committee, 'Preliminary Report', 12.5.16.
8. POWE 33/41 Conf. de Bunsen to Admiralty, 3.3.16, in F. Parker, Conf., 'Memorandum respecting oil concessions in Mesopotamia', 27.4.18, Annex 2.
9. *Ibid.*
10. POWE 33/2 Private. Memorandum. Lord Cowdray to Sir Francis Hopwood, 12.8.16.
11. *Hansard*, 1916, vol. 80, W. Runciman, 15.5.16.
12. CAB 27/180/C.P.4050/Secret. Committee on the Proposed Amalgamation of the *Royal Dutch-Shell, Burmah* and *Anglo-Persian Oil Co.*, 'Report', 22.6.22.
13. *Hansard*, 1916, Vol. 85, Major Hunt, 8.8.16.
14. *Financial Times*, 4.12.17.
15. Royal commission on Dominions, op. cit.
16. Ibid., p.71.
17. POWE 33/1 Sir John Cadman to Walter Long 28.5.17. In Britain the Mineral Oil Production Department had supreme control over oil production in the country from shale, coal, and oil. The Pool Board undertook distribution of oil within the British Isles, while the Petrol Control Department, staffed by over 300 people, existed to keep a check on oil consumption. The Petroleum Executive co-ordinated all oil supplies and became in 1920 the Petroleum Department under the Secretary for Overseas Trade. Later in October 1922 it was absorbed into the Board of Trade becoming in 1928 a branch of the Mines Department of the Board of Trade. In 1934 there were only three full-time officials in the Department, increasing to thirteen in 1938. In France the Commissariat General for Petrol and Oil Fuel was set up in 1917 to supplement the Petrol Import Syndicate. (cf. Anton Mohr, *The Oil War* (London: Martin Hopkinson & Co. Ltd., 1926), and D.J. Payton Smith, *Oil. A Study of War-Time Policy and Administration* (London: HMSO, 1971).
18. In late 1916 the *Royal Dutch Co.* arranged through the New York broking house of *Kuhn, Loeb & Co.* to sell 74,000 *Royal Dutch* shares on the American market. Receipts were between $12-13 million, and part of this money was used to purchase material and equipment in the U.S. 'for shipment abroad to refineries then under construction in Egypt, Venezuela and Curaçao'. (Kendall Beaton, *Enterprise in oil. A history of Shell in the United States* (New York: Apple-Century Crofts Inc., 1957), p.167.
19. POWE 33/4 Sir Marcus Samuel to Long, 1.8.17.
20. POWE 33/3 Samuel to Long, 12.7.17. A further attraction which Venezuela held, which was pointed out as early as 1915, was that it would not be subjected to military attacks, and consequently the Navy would be guaranteed its supplies. The Turkish attack on Ahwaz, for example, had as its real objective *Anglo-Persian*'s pipeline. The *Financier* wrote:

> The protection of the pipeline must be a matter of constant anxiety, and it is therefore of importance to this country to encourage the production of oil in parts of the world where the enemy has no power to interfere with the supply . . . In Venezuela, for instance, there is a clear field and there the Venezuelan Oil concessions possess a property covering an area of 3,000 square miles. (The *Financier*, 30.4.15)

21. MUN 5/215/1970/17 Telegram, Sir C. Spring-Rice to Foreign Office, 10.10.17.
22. Ministry of Munitions, 'Copy of an agreement made between the Minister of Munitions and S. Pearson & Sons Limited on 10 September 1918 for Management of Petroleum Development', *PP* 1918 Vol. XV Cmd. 9188, pp.743-50, Lord Cowdray to Editor of *Hansard*, 1.3.18, pp.748-9.
23. *Hansard,* 1917, Vol. XCV, Dr. C. Addison, 28.6.17.
24. *Hansard,* 1917, Vol. XCVII, W. Long, 15.8.17.
25. *Ibid.*
26. *Hansard,* 1918, Vol. 103, 27.2.18.
27. POWE 33/44 'Oil Policy of the Empire', Undated. The position of the U.S. and the Allies was the more serious because for the duration of the war the U.S. consumed oil 'in excess of production, the difference being covered by imports from Mexico, supplemented later by drafts from domestic stocks'. (Mark L. Requa, 'Report of the Oil division, 1917–1919', in H.A. Garfield, *Final Report of the U.S. Fuel Administration, 1917–1919* (Washington: USGPO, 1919), pp.259-305, p.277). By 1916 President Wilson's Administration had ordered a close check on oil consumption, and, under the Council of National Defence, the Petroleum Advisory Committee (representing the petroleum industry) was formed with Alfred C. Bedford as Chairman. When the U.S. entered the War in 1917 the Oil Division headed by Mark. L. Requa of the U.S. Fuel Administration was formed to ensure that the Army, the Navy shipping board, and the armaments industry, as well as domestic consumers were allocated oil fairly. Under the stimulus of war conditions the price of crude oil in 1918 doubled its pre-war level. Prices were nevertheless stabilised by the combined efforts of the National Petroleum War Services Committee (the old Petroleum Advisory Committee reorganized in 1918) and the Oil Division of the U.S. Fuel Administration. On August 9, 1918, a voluntary agreement known as 'The Plan' was arrived at between the industry and government in order to maintain the flow of oil and stabilise prices (cf. Joseph E. Pogue, *Prices of petroleum and its products during the war* (Washington: USGPO, 1919).
28. Mexican, Russian and Romanian oil supplies were to contribute little in the future owing to a projected increase in domestic demand that would leave smaller surpluses for export.
29. POWE 33/34 'Oil Policy of the Empire', Undated.
30. POWE 33/2 'Progress of the negotiations with regard to the Burmah-Shell Amalgamation', Undated.
31. POWE 33/44 'Oil Policy of the Empire', Undated; POWE 33/42 Conf., 'On the Policy to be adopted by His Majesty's Government with regard to the proposals which have appeared in the Press for the Formation of an All-British Oil Company at the Colonial Office', 12.12.17; and, *Hansard,* 1917, Vol. C, Col. 841, Sir J.D. Rees, 10.12.17.
32. *Hansard,* 1917, Vol. C, Col. 841.
33. POWE 33/42 Press Notice, Petroleum Department, 1.1.18.
34. POWE 33/42 Cadman to Batterbee, 2.1.18.
35. POWE 33/42 Cadman to Batterbee, 12.12.17.
36. POWE 33/42 Samuel to Cadman, 12.10.17.
37. *Ibid.*
38. *Ibid.*
39. POWE 33/42 Cadman to Batterbee, 11.12.17.
40. POWE 33/2 'The future control of oil supplies', Undated.
41. POWE 33/42 Long to Sir Albert Stanley, 29.12.17.
42. POWE 33/8 Henry Bérenger, 'Note read at the Fourth Preliminary Meeting of the Inter-Allied Petroleum Conference', 27.2.18.

43. POWE 33/12 4.3.18. The terms of reference were changed in September to 'ensure the control and supply of oil' (POWE 33/13 Petroleum Executive to Batterbee, 25.9.18).
44. POWE 33/12 Cadman, 'Memorandum', 25.5.18.
45. *Ibid.*
46. POWE 33/13 Petroleum Imperial Policy Committee, First Meeting, 29.5.18, 'Introduction Memorandum'.
47. POWE 33/13 'Minutes of Second Meeting held at the Offices of H.M. Petroleum Executive', 19.6.18.
48. *Ibid.*
49. POWE 33/13 'Notes of a Meeting of Lord Harcourt, Sir John Cadman, Sir Henry MacGowan, and Mr. Lancelot Smith, representing the Policy Committee, and Sir Marcus Samuel and Mr. H.W.A. Deterding on behalf of the *Royal Dutch-Shell*', 10.7.18.
50. *Ibid.*
51. The 'Shell' Transport & Trading Co. Ltd., *Annual Report*, 1918.
52. *Hansard,* 1919, Vol. 119, 18.8.19. The shares were later sold abroad and the proceeds used to support the exchange rate between Britain and Holland.
53. POWE 33/13 'Imperial War Conference', 22.7.18.
54. Never published.
55. Imperial War Conference, 1918, 'Extracts from Minutes of Proceedings and Papers laid before the Conference', *PP* Vol. 16, 1918, Cmd. 9177, pp.691-942, Resolution 18, 22.7.18, p.697.
56. Petroleum Executive, 'Report of the Inter-Departmental Committee on the Employment of Gas as a source of Power, especially in motor vehicles in substitution for petrol and petroleum products', *PP*, 1919, Vol. 22, Cmd. 263, pp.521-68. For the Committee's conclusion see p.26.
57. H.M. Petroleum Executive, 'Report of the Inter-Departmental Committee on various matters concerning the production and utilisation of alcohol for power and traction purposes', *PP*, 1919, Vol. 10, Cmd. 218, pp.117-24.
58. Ministry of Munitions of War, 'Report of a Committee appointed by the Right Hon. The Minister of Munitions respecting the production of fuel oil from home sources', *PP,* 1918, Vol. X, Cmd. 9128, pp.515-22.
59. Geo Cave to Lord Cowdray, 20.3.18, in Ministry of Munitions, 'Agreement with S. Pearson', *op. cit.*, pp.750.
60. Geo Cave to Lord Cowdray, 12.3.18, in Ministry of Munitions, 'Agreement with S. Pearson', *op. cit.*, pp.749.
61. *Hansard,* 1918, Vol. 109, Col. 238.
62. *Ibid.*
63. *Journals of the House of Commons,* Vol. 173.
64. Any landowner would be granted a licence provided the requirements of the Committee of Geology were fulfilled.
65. Ministry of Munitions, 'Agreement with S. Pearson', *op. cit.*
66. *Public General Acts,* 1918.
67. The Admiralty felt that 'the gradual substitution of Oil for Coal will in the future, if the question is not immediately faced, wrest from our grasp one of the principal factors on which the maintenance of our Naval position depends' (POWE 33/45 'Petroleum situation in the British Empire. Admiralty Memorandum for the Imperial War Cabinet', Undated).
68. CAB 21/119 Admiral Sir Edmond Slade, 'Petroleum situation in the British Empire', 29.7.18.
69. CAB 21/119 M.P.A. Hankey to Sir Eric Geddes (Admiralty), 30.7.18.
70. CAB 21/119 Hankey to Prime Minister, 'Memorandum', 1.8.18.
71. CAB 21/119 Hankey to Geddes (E.), 30.7.18. In 1917 the Admiralty had

already voiced its opinion when it felt that

> it would appear . . . highly desirable that German economic concessions in such countries as Finland, Estonia, Livonia, Ukraine, Turkey, Roumania, Bulgaria and the Caucasus etc. should be if possible cancelled in order that Allied enterprises may have opportunities of securing what will be in many cases valuable properties. (ADM 137/2853 Naval Staff, Trade Division, 'Memorandum', 30.12.17.)

72. ADM 137/2826 Additional Naval Assistant to First Sea Lord, Director of Trade Division, Admiralty, 'Oil Fuel Supply-Memorandum from the Strategic Point of View', 26.9.18.
73. POWE 33/13 Slade, 'Memorandum on the Petroleum Situation in the British Empire', 9.10.18.
74. ADM 137/2826 Additional Naval Assistant to First Sea Lord, *op. cit.*
75. POWE 33/13 Petroleum Executive to Batterbee, 25.9.18.
76. POWE 33/13 Minutes of 9th Meeting of H.M. Petroleum Executive, 21.10.18.
77. POWE 33/13 Lord Inchcape to Lord Harcourt, 17.10.18 in 'Minutes. 9th Meeting of H.M. Petroleum Executive', 21.10.18.
78. POWE 33/13 Sir Harry MacGowan to Lord Harcourt, 31.10.18, in 'Minutes. 11th Meeting of H.M. Petroleum Executive', 1.11.18.
79. POWE 33/13 'Minutes. 12th Meeting of H.M. Petroleum Executive', 11.11.18.
80. POWE 33/13 'Minutes. 13th Meeting of H.M. Petroleum Executive', 14.11.18.
81. cf. Francis Delaisi, *Oil, its influence on politics* (London: The Labour Publishing Co. Ltd. & George Allen & Unwin, 1922).
82. In the U.S. there was a phenomenal increase in motor vehicles. The number of cars increased from 1.6 million in 1914 to 5.6 million in 1918 to 27.4 million in 1940. The number of trucks also increased from 85 thousand in 1914 to 525 thousand in 1918 to 4.6 million in 1940. (cf. John W. Frey & H. Chandler Ide, *A History of the Petroleum Administration for War, 1941-45* (Washington: US GPO, 1946).
83. POWE 33/60 Cadman, 'Petroleum Position of the British Empire', December 1918.
84. 'Report on Motor Fuel prepared by a Sub-Committee appointed by the Standing Committee on Investigation of Prices', *PP* 1919, Vo. 23, Cmd. 597, pp.573-82.
85. H.M. Petroleum Executive, 'Report of the Inter-Departmental Committee on the Employment of Gas as a source of power, especially in motor vehicles in substitution for petrol and petroleum products'. *PP* 1919, Vol. 22, Cmd. 263, pp.521-68.
86. POWE 33/13 'Minutes. 13th Meeting of H.M. Petroleum Executive', 14.11.18.
87. POWE 33/13 'Minutes. 15th Meeting of H.M. Petroleum Executive', 26.11.18.
88. POWE 33/13 'Minutes. 16th Meeting of H.M. Petroleum Executive', 5.12.18.
89. CAB 21/119 E. Montagu (Secretary of State for India) to Lord Harcourt, 23.12.18.
90. *Ibid.*
91. *Ibid.* Montagu also suggested that in order to safeguard the interests of the new State it should be allowed to participate in the new company.
92. POWE 33/13 H.M. Petroleum Executive, 'Negotiations regarding the Petroleum Policy of His Majesty's Government', Vol. 1, 'Report and

Proceedings of the Petroleum Imperial Policy Committee', 29.5.18 to 10.2.19.
93. POWE 33/71 'Memorandum', 27.2.19.
94. POWE 33/13 'Minutes. 18th Meeting of H.M. Petroleum Executive', 20.1.19.
95. POWE 33/13 'Minutes. 20th Meeting of H.M. Petroleum Executive', 3.2.19;
and, POWE 33/68 'Memorandum – Initialled by Lord Harcourt and Henri
Deterding', 6.3.19. The Petroleum Imperial Policy Committee ceased to
function on 10 February 1919.
96. POWE 33/71 'Memorandum. *Shell* and *Royal Dutch* Income Tax Provision',
8.5.19.
97. POWE 33/68 'Memorandum for the War Cabinet (Acquisition of British
Control over *Royal Dutch-Shell* Group)', Undated.
98. POWE 33/68 Letter, Captain A.S. Jelf (Petroleum Executive) to J.
Schuckburgh (India Office), London, 4.6.19. The Petroleum Department in a
Memorandum for the War Cabinet argued that:
These arrangements secure a very important measure of permanent British
control over this very powerful group and although for the present no
conspicuous change in the personnel or policy can be looked for, or could
perhaps be insisted on without loss of efficiency, British influence in the
group is greatly strengthened and ultimate predominance is secure.
(POWE 33/68 Petroleum Department, 'Memorandum for the War
Cabinet (Acquisition of British Control over the Royal Dutch-Shell
Group)', Undated).
99. POWE 33/71 A. Stuart & H. Colijn (Shell Reps.) to Cadman, 28.11.19.
100. *Ibid.*
101. *Ibid.*
102. *Ibid.*
103. *Ibid.*
104. POWE 33/71 Harold Brown (Linklaters & Paines) to Cadman, 8.12.19.
105. POWE 33/71 H.M. Petroleum Executive, 'Note of Interview held at the
Offices of the Asiatic Petroleum Company on Thursday the 16th December,
1919.'
While this occurred the government's holdings in *Anglo-Persian* increased
from £2 million to £5 million. The Treasury viewed this as a purely
commercial transaction, but as Sir John Cadman pointed out, it had serious
'disadvantages and handicaps us not only in our dealings with other British
companies but also with foreign governments'. (POWE 33/67 Cadman to Sir
Harvey Greenwood, 4.12.19). On 23 January 1920, the Cabinet decided that as
a matter of principle the revenues arising from the exploitation of the
Mesopotamian oilfields would accrue to the State formed and would not go to
the joint company formed to exploit them. (POWE 33/67 W.S.C[hurchill],
'Mesopotamian Oil and the Turkish Petroleum Company. Memorandum by
the Secretary of State for the Colonies', 29.8.21).
106. POWE 33/71 'A meeting in regard to the Royal Dutch Shell Agreement took
place on Thursday 26 March (1920)'.
107. POWE 33/71 Brown to Cadman, 14.4.20.
108. *Ibid.*
109. The question of *Anglo-Persian*'s pipeline running through French territory
was also touched on during these talks.
110. POWE 33/14 'Anglo-French Agreement', Undated.
111. The same day when the Lord Harcourt-Deterding Agreement was initialled.
112. POWE 33/176 Petroleum Department, 'Memorandum for the War Cabinet',
9.12.19.
113. POWE 33/176 'Anglo-French Agreement', Initialled W.H.P., 9.1.20.
114. *Ibid.*

115. This was a Peace Treaty between Turkey and the Allied Powers. The same Treaty recognised that France would rule Syria as a League of Nations Mandate.

116. Memorandum of Agreement between M. Philippe Berthelot, Directeur des Affaires Politiques et Commerciales au Ministère des Affaires Etrangères and Professor Sir John Cadman K.C.M.G., Director in Charge of His Majesty's Petroleum Department', *PP* 1920 Vol. LI, Cmd. 675, pp.895-8.

117. One of President Wilson's Fourteen Points for Peace sent to the Germans on 8 January 1918, was the removal of all economic barriers and the establishment of equal trade conditions. The San Remo Agreement was seen, therefore, as a way to prevent U.S. economic expansion.

118. Shwadran, *op. cit.*, p.204.

119. POWE 33/176 Telegram 811 Geddes (A.) to Lord Curzon, 16.12.20.

120. Up to March 1920 he had been President of the Board of Trade, and therefore was well briefed on the various petroleum negotiations which had taken place.

121. POWE 33/176 Telegram 619, Lord Curzon to Geddes (A.), 23.7.20.

122. According to Shwadran this was done by paying the Shah a monthly subsidy of 15,000 'tomans'.

123. Other Anglo-Persian enterprises such as the construction of railways were to be encouraged.

124. Davenport & Cooke, *op. cit.*, p.46. Venezuela was one of these countries.

125. Mikdashi states that between 1914 and 1924 the non-consolidated profits of *Anglo-Persian* when only handling Persian oil were £28.5 million, of which the shareholders received £9.5 million and the Persian government £3.9 million, or almost 13.7 per cent. (Cf. Zuhayr Mikdashi, *A Financial Analysis of Middle Eastern Oil Concessions: 1901-65* (New York: Frederick A. Praeger Publishers, 1966).

126. In July 1918 the company with a capital of £250,000 was formed to drill for oil in the UK, obtaining the following year a two year licence to bore and exploit for oil and gas in Newark, Nottinghamshire. Its licence was extended for a further two years in May 1921. ('Petroleum Production (Licences). Licence dated 12 May 1919, granted by the Minister of Munitions to Oilfields of England, Limited' *PP* 1919, Vo. XLII, Cmd. 1917, pp.1047-52); 'Rt. Hon. William C. Bridgeman M.P. (Secretary for Mines) and Oilfields of England Limited Licence', *PP* 1921, Vol. XXXI, Cmd. 1434, pp.541-6; and, *Skinner's Oil Manual*, 1919, p.107.

127. In October 1920 Reginald Gilbey obtained a licence to drill for oil over 335 acres in Leicestershire. ('The Minister of Munitions and Reginald Gilbey. Copy. Licence as to boring for petroleum at Weston, in the county of Leicester', *PP* 1920, Vol. 30, Cmd. 1031, pp.257-62).

128. *Hansard,* 1919. Vol. 116, Col. 1423.

129. *Hansard,* 1919 Vol. 116, Statement by Mr. Kellaway, Deputy Minister of Munitions. The Duke of Devonshire, who owned the land on which oil was discovered, was unwilling to give up his rights to the oil produced, and eventually acquired the licence in March 1923. The government in November 1919 decided that no royalty would be payable on oil. (Cf. *Hansard,* 1919, Vol. 118, 24.6.19; *Ibid.,* 1919, Vol. 120, Col. 24: and, 'Licence to drill for Petroleum granted by the Secretary of State for Mines to His Grace the Duke of Devonshire', *PP,* Vol. 19, 1923, Cmd. 1873, pp.659-66).

130. *Hansard,* 1920, Vol. 135, 7.12.20 James Hope to Secretary to Minister of Munitions.

131. 'Profiteering Acts, 1919 & 1920. Second Report on Motor Fuel prepared by a Sub-Committee appointed by the Standing Committee on the Investigation of Prices', *PP* 1921, Vol. 6, Cmd. 1119, pp.793-804.

132. *Ibid.*, p.800.
133. *Ibid.*, p.800.
134. *Ibid.*, p.801.
135. POWE 33/176 Sir R.S. Horne (Cabinet), 'Mesopotamian Oilfields. Memorandum by the President of the Board of Trade', 16.4.20.
136. *Ibid.*
137. *Ibid.*
138. POWE 33/92 J.C. Clarke, 'Note to Sir Philip Lloyd-Greame', 11.10.21, Enclosure, 'Re: Proposed Trading Agreement for the sale of Benzine in the United Kingdom and Ireland', 6.7.21.
139. Later Lord Swinton.
140. POWE 33/92 'Notes of Meeting held at the President's Room at the Board of Trade, 27 July 1921, on Proposals for Joint Distribution of Petroleum by the *Anglo-Persian Oil Company* and the *Shell* Group'.
141. *Ibid.*
142. POWE 33/92 Clarke, 'Note to Sir Philip Lloyd-Greame', 11.10.21.
143. POWE 33/92 'Proposed Combination of the Royal Dutch Shell, Burmah and Anglo-Persian Oil Companies put forward by Burmah Oil Co., 29 July 1921'. Enclosed in Clarke to W.St.D. Jenkins, 11.10.21.
144. The Admiralty wanted the new Group to be formed to supply it with 250,000 tons of fuel oil annually (to be increased in the future to 400,000 tons) at £1 per ton and the quality to be approved by the Admiralty. They also stipulated a right to terminate the contract after notice of two years (POWE 33/92 'Summary of Memorandum prepared by the Director of Contracts, Admiralty, in regard to the proposed amalgamation of Royal Dutch-Shell, Burmah and Anglo-Persian Oil Company', Undated).
145. POWE 33/92 'Proposed combination of Royal Dutch-Shell, Burmah and Anglo-Persian Oil Company – Notes of meeting held in Sir Philip Lloyd-Greame's Room on Wednesday 26 October 1921'.
146. *Ibid.*
147. It should be remembered that the Dutch shareholders held a majority interest in all *Shell* companies and could whenever they desired transfer the lot to Holland and in an extreme case sell to *Exxon*.
148. POWE 33/92 Robert Watson, 'As to permanent and effective British control of the Group or Groups referred to in the memorandum of 29 July 1921'. Undated.
149. POWE 33/92 'Memorandum prepared by the Petroleum Department', 9.11.21.
150. The British government gained considerably from its original investment of £2 million in the company. Up to 1922 it benefitted by nearly £22 million, divided as follows:

Dividends, interest, income tax etc.	£5,360,000
Gain on Admiralty and War Office contracts as compared to what they would have had to pay	£7,050,000
Gain on purchase of oil from other suppliers due to Anglo-Persian's competition (hypothetical figure)	£5,000,000
Share appreciation	£4,500,000
TOTAL	£21,910,000

This does not take into account the government's later investment in 1919 of £3 million. The gains from this were calculated as being slightly over £17 million, divided as follows:

Appreciation in value of £3 million ordinary shares subsequently acquired	£6,750,000
13 years deliveries of Admiralty fuel oil contract still left to run, 5 million tons at £2 per ton	£10,400,000
TOTAL	£17,150,000

Source: CAB 27/180/Secret/O.S.C.4 'Cabinet Committee on Oil Amalgamation – Estimates of return obtained by His Majesty's Government on their original investment of £2,000,000 in the Anglo-Persian Oil Company', 23.1.22.

151. POWE 33/92 'Memorandum prepared by the Petroleum Department', 9.11.21.
152. *Ibid.*
153. This was illegal as under the Venezuelan oil laws of 1920, 1921, 1922, 1925, and 1928 no company however remotely connected with a foreign government was allowed to hold concessions.
154. *Skinner's Oil Manual*, 1923.
155. U.S. National Archives, Department of State (DS) 831.6363/59 Conf. Dudley G. Dwyne to Preston McGoodwin, Maracaibo, 4.5.21.
156. cf. DS 831.6363/Lago Pet. Corp/1 Fred H. Kay, 'History of Petroleum Concessions owned by Lago Petroleum Corporation in Venezuela', April 1928. In 1924 the *Lago Pet. Corp.* acquired the company for $2 million (DS 831.6363/L13/Org. Robert P. Skinner, 'A Venezuelan Oil Deal', 14.3.24).
157. *Skinner's Oil Manual*, 1920, p.123.
158. FO 371/8493 G. Armstrong to Geddes (A), 16.12.22.
159. POWE 33/92 Clarke to Lloyd-Greame, 15.12.21.
160. *Ibid.*
161. POWE 33/92 Lloyd-Greame, 'Proposed amalgamation of the Royal Dutch-Shell, Burmah and Anglo Persian Oil Company', 6.1.22.
162. *Ibid.*
163. POWE 33/92 Watson to Clarke, 31.1.22.
164. *Ibid.*
165. POWE 33/92 T. St. Quintin to Lloyd-Greame, 2.2.22. Sir Philip was to serve on the Committee.
166. Russian oil men had close links with France. The *Neft* and the *Barkinski*, two of the largest Russian oil companies, were quoted on the Paris Bourse. After the 1917 Revolution and the nationalisation of the oil companies, many Russians settled in Paris where they obtained cash advances from certain Paris banks by pledging their confiscated properties. *Exxon* after the war acquired a 50 per cent stake in *Nobel*'s Russian oil properties, which controlled 40 per cent of the pre-war Baku oil production. It was fully expected that the oil companies would be allowed to remain in business after the war. The Bolshevik government was not expected to last and it was not clear whether the oil region of the Caucasus would remain in Russian hands. After the military collapse of the Russians, the Turks launched an attack on Kars, Batum, and subsequently on Baku to secure the rich oilfields. A Trans-Caucasian Federation consisting of the Republics of Georgia, Armenia, and Azerbazjhan was set up. But, at the request of the Georgians, the Germans in May 1918 occupied the region and the Federation was dissolved. Later the British occupied the area for a brief spell, but were driven back to Persia by the Turks, who managed to denationalise the oilfields and get the oil flowing again. Their stay was short lived, and with the signing of the Armistice on 11 November, the British re-entered Baku prepared for a long stay. The Baku–Batum pipeline was

overhauled, and essential services established. Despite this the Bolsheviks entered Baku in April 1922, and a year later Georgia became a Bolshevik republic. (Cf. Davenport & Cooke, *op. cit.*; Louis Fischer, *Oil Imperialism* (London: George Allen & Unwin, 1926); Nikpay, *op. cit.*; and Shwadran, *op. cit*).

167. cf. Fischer, *op. cit.*, p.44.
168. Moreover, *Shell* and *Exxon* refused to co-operate. Deterding had previously tried to reach a private agreement with the Russians.
169. According to Taylor the Conference was convened to 'put every great problem to rights. German reparations were to be settled. Soviet Russia was to be brought back into the comity of nations, and her markets reopened to international trade. The United States were to write off the war debts of the Allies' (A.J.P. Taylor, *English History, 1914-1945* (Oxford: Oxford University Press, 1965), p.189).
170. Fischer, *op. cit.*, p.93.
171. CAB 27/180/Secret/O.S.C.2 'Conference. Cabinet Committee on Oil Amalgamation – Conclusions of a meeting of the above Committee held in the Minister's Conference Room', 13.3.22.
172. *Ibid.*
173. FO 371/7279 Telegram 219 Secret, Geddes (A.) to Lord Curzon, 4.5.22. It should be noted that Sir Auckland as President of the Board of Trade between 1919-20 was directly involved with the merger negotiations.
174. *Ibid.*
175. CAB 27/180/C.P.4050 Secret. Committee on the Proposed Amalgamation of the Royal Dutch Shell, Burmah and Anglo Persian Oil Company, 'Report', 12.6.22.
176. *Ibid.*
177. *Ibid.*
178. Confidential, Maurice de Bunsen to Admiralty, London, 3.3.16 in F. Parker, 'Confidential. Memorandum respecting oil concessions in Mesopotamia, 27 April 1918, Annex 2', POWE 33/41.
179. For example, Iraq only started to export oil in large quantities in 1934 when the pipeline to the Mediterranean was completed.

3

Anglo American Oil Diplomacy

At the end of the First World War the U.S. was overcome by a veritable oil scare with the British bearing the brunt of their frustration.[1] It had been freely predicted in the U.S. that within a short period the country would be 'absolutely dependent on foreign countries, especially on the British Empire and other sources under British supplies, for oil and consequently for her existence as a powerful industrial and maritime nation'.[2] Indeed, wherever *Exxon* looked it saw the 'Shell' trade mark, and as Sir Edgar MacKay of *Sperlings & Co.* (merchant bank) stated 'it was almost a case of the British first and the rest of the world nowhere'.[3]

In 1919, David White, Chief Geologist at the U.S. Geological Survey, estimated the country's oil reserves at 6.7 billion barrels, and foresaw exhaustion of these reserves in 17 years' time. By 1925 it was predicted that the U.S. would rely on oil imports to the extent of 150 million barrels per annum.[4] On 12 May 1919, the General Director of the Oil Division of the U.S. Fuel Administration urged that 'in the national interest, American petroleum companies be encouraged by the Government to acquire foreign sources of oil supplies, wherever they can be obtained'.[5] A joint report submitted on 16 May 1919, by the Fuel Administration, State Department, Bureau of Mines, and Geological Survey, recommended that the State Department 'should exert every effort to encourage and protect its citizens in securing petroleum wells and concessions for petroleum development in foreign countries and in working these properties and concessions after they have secured them'.[6] Secretary of the Interior Wilbur wanted to place the entire U.S. oil industry under conservation and use foreign oil.[7] By 1920 a serious oil shortage existed in the U.S., and despite a domestic increase of oil production of 22 million barrels and imports amounting to 51 million barrels, oil stocks diminished by 15 million barrels.[8] There was fierce competition among the 292 'oil refiners for the material with which to keep their plants working up to an economical percentage'.[9] As nearly two-thirds of these were owned by *Exxon* companies or the sixteen largest independent refiners[10] the shortage led several *Exxon* companies to amend their charters to enable them to engage in crude oil production as well as refining; for example, *Exxon*'s Foreign Production Department was established immediately after the war.[11] The Federal Trade Commission in a report presented in 1920 recommended, once again, that oil companies developing foreign oilfields should be 'given all diplomatic support in obtaining and operating oil producing property'.[12]

The U.S. had always been able to supply herself and her foreign markets from her own oil resources. However, her now depleted oil stocks raised the possibility of her having to rely in the future on foreign (mainly British) controlled oilfields situated in the Middle East and in the British Empire. The competition for the world's oil supplies took on a decidedly intense turn, and especially in the Middle East and Venezuela, where it was so keen, some even felt that Britain and the U.S. were on the verge of declaring war.[13] The U.S. emerged after the war with greater prestige, demanding and succeeding in getting the world to recognise her as an emerging world power. Up to then, the Caribbean and Central America, and to varying degrees the rest of Latin America, had been recognised as American spheres of influence. The U.S. with its stronger post-war economy demanded from the Allies equal participation and fair competition in the world's markets. Translated into oil matters this was the adamant demand for an 'open door' policy to be pursued in all Allied colonies and mandated Territories.[14]

During this time a number of Senators and Congressmen were critical of the role played by the British government in trying to monopolise the world's oil resources. The Polk Senate report of 1919, for instance, severely criticized Britain's policy towards foreign oil companies exploiting the oilfields of her Colonies and Mandated Territories, especially Mesopotamia. On the same day this report appeared, Senator Phelan of California introduced a Bill in Congress requesting the creation of a U.S. government owned Oil Corporation which would exploit foreign oil resources. Secretary of the Navy Josephus Daniels was so worried by the oil shortage that he was willing to commandeer oil supplies under the broad powers of the Lever Act. In a speech given to the American Society of Naval Engineers, Daniels suggested that all U.S. petroleum resources be nationalised in order to protect future needs of the Navy.[15] In 1920 Senator Smoot introduced the Mineral Oil Leasing Bill, which Congress subsequently approved, and which stated *inter alia* that so far as Federal lands were concerned concessions could only be awarded to foreign citizens or foreign companies of countries which had reciprocal oil agreements with the U.S. Holland and Britain did not fall within these criteria[16] as they did not allow foreign companies to exploit the oil resources in their Colonies. The Senators were almost unanimous in their agreement that safeguards had to be erected against the exploitation by foreigners of U.S. oil in publicly held land. This would be a severe blow to foreign oil companies especially to the *Shell* Group which obtained one-third of its supplies from the U.S., as private land in known oilbearing regions was becoming more expensive to obtain. Senator Phelan was however unsuccessful in his efforts to effect a total ban of aliens from public land.

Nevertheless, under President Harding, Secretary of the Interior Albert Fall refused the *Roxana Petroleum Co.* (a *Shell* subsidiary) application for oil leases on Creek Indian land in Oklahoma State.[17] The individual states were also encouraged to enact similar state legislation to that of the Leasing Act. Senator Ferral introduced a Bill in the Oklahoma Legislature which would oblige *Shell* or any other foreign company to dispose of all their leases in the State to U.S. citizens. Similar legislation was enacted in Montana and Wyoming, while in Texas the law made it impossible for a British controlled company to own land.[18] These attacks had the clear objective of inducing foreign governments, especially Britain's, to give more favourable treatment to U.S. companies in foreign oilfields from which they were excluded.[19]

Exxon also echoed the U.S. government's desire to be allowed to develop foreign oilfields under British control. Walter C. Teagle, President of the company, in one of the opening speeches of the Annual American Petroleum Institute's Convention held in November 1920, reaffirmed the State Department's view that foreign governments were 'deliberately placing obstacles in the way of those who would like to assist in the development of new sources of supply'.[20] Teagle added that foreign oil was needed to supplement dwindling domestic supplies. While in 1921 oil imports from Mexico reached a record level of 121 million barrels,[21] it was apparent to the companies that Mexican production was faltering due chiefly to technical problems associated with waterlogging.[22] In addition the unstable political situation there indicated the need to find a secure alternative source of oil to supplement U.S. domestic oil production. Though Venezuelan production would later fill this gap, the U.S. government's attention was focused firmly on events happening in the Middle East and Britain's role there. The U.S., however, was far from running out of oil, and would witness during the 1920's, starting with the Texas Panhandle discovery, a wave of major oil finds in Oklahoma, California and Wyoming. By 1921, there was such a glut of oil on the market that the Independent producers tried to induce Congress to impose a tariff on cheap foreign oil, mainly Mexican, as their profit margins were rapidly diminishing. President Harding opposed such a proposal on the grounds that the future military and domestic needs made it desirable for the U.S. to exploit foreign oil sources; moreover, an oil tariff would be against the interests of American companies seeking foreign oil because it was feared that a tariff would stop companies from investing abroad.[24]

During the interim period of military occupation the U.S. government had exchanged a number of Notes with the British government over the administration of the Mandated Territories of Palestine and Mesopotamia. The U.S. contended that British oil interests had been given favourable advantages over American companies. This view stemmed from the belief that pipelines, railways, refineries and dockyards had been constructed in the region.[25] Lord Curzon contested this view in affirming to the U.S., that no pipelines or refineries had been constructed in Mesopotamia. In 1920 a start had been made on a small refinery for military needs in Baghdad which was 'intended to deal with oil obtained from the Persian oilfields'.[26] Lord Curzon further repudiated the U.S. government's suggestion that during the period of military occupation the UK had prepared the way for British companies only to exploit the oilfields of the region. Lord Curzon informed John Davis, American Ambassador at London, that the claims of British companies 'are today no stronger, as they are no weaker, than they were at the outbreak of war'.[27] The allegation that British oil interests wanted to bar American companies from developing the world's oil resources was 'singularly unintelligible'.[28] The facts, as Lord Curzon interpreted them, were that the U.S. produced nearly 70 per cent of the world's oil and that American companies operating in Mexico accounted for a further 12 per cent of the world's total. In contrast the the Empire's meagre output of 2.5 per cent of world oil production compared very unfavourably, demonstrating that British oil companies were not retaining the oilbearing lands of the Empire for their own use. Moreover, the U.S. government, through the enactment of the Mineral Leasing Act, had taken 'powers to reserve for American interests the right to drill for oil on United States domain lands and have on various occasions used their influence in territories amenable to their control with a view to secure the cancellation of oil concessions previously and legitimately obtained by British persons or companies'.[29] Despite these arguments the U.S. government believed

> that it is entitled to participate in any discussion relating to the status of such concessions, not only because of existing vested rights of American citizens, but also because the equitable treatment of such concessions is essential to the initiation and application of the general principles in which the United States Government is interested.[30]

The U.S. government was aware that the administration of the mandate would involve heavy financial obligations to the UK but that any 'reimbursement by the adoption of a policy of monopolisation or of exclusive concessions and special favours to its own nationals, besides being a repudiation of the principles already agreed to, would prove to be unwise from the point of view of expediency both on economic and political grounds'.[31] Up to the time of the signing of the Versailles Peace Treaty the U.S. government

> consistently took the position that the future peace of the world required that, as a general principle, any alien territory which would be acquired pursuant to the Treaties of Peace with the Central Powers must be held and governed in such a way as to assure equal treatment in law and in fact to the commerce of all nationals. It was on account of, and subject to this understanding that the United States felt itself able and willing to agree that the acquisition of certain enemy territory by the victorious Powers would be consistent with the best interests of the world. The representatives of the principal Allied Powers in the discussion of the mandate principle expressed in no indefinite manner their recognition of the justice and far-sightedness of such a principle and agreed to its application to the mandates over Turkish territory.[32]

It was therefore felt that friendly co-operation between the citizens of the U.S. and the UK would be in the best interests of both countries. In American eyes the San Remo Agreement resulted in a 'grave infringement of the mandate principle which was formulated for the purpose of removing in the future some of the principal causes of international differences'.[33]

In the U.S., Senator Phelan continued his anti-British attacks. In San Francisco he stated that the British were scheming 'desperately to control the oil supply of the world, to the detriment of the United States'.[34] The American Petroleum Institute, formed in 1919 with the former Directors of the National Petroleum War Services Committee, discussed these issues at its annual convention, held November 18-21. Sir Auckland Geddes felt it unwise for Sir John Cadman as representative of the British government to address the meeting, as the whole convention was part of a 'scheme for extending American control over the world's oil supplies'.[35] W.C. Teagle, President of *Exxon*, in one of the opening speeches reaffirmed the State Department's view that foreign governments were 'deliberately placing obstacles in the way of those who would like to assist in the development of new sources of supply'.[36] Teagle added that foreign oil was needed to supplement dwindling domestic supplies. Otis Smith, Director of the U.S. Geological Survey, confirmed the view held by the President of *Exxon* that American stocks would 'last only nine years and three months',[37] and declared that there was

> an urgent need of pioneering the world for oil to meet the needs of this generation, but there is no warrant for regarding this advance into new fields as beginning a contest whose aim is world conquest. The present need of the United States for oil from abroad can be met only by world wide exploration, development and operation by American companies backed by our government.[38]

Richard Airey, Vice President of *Roxana Petroleum*, voiced the European opinion when he stated that Europe needed to acquire its own oil supplies rather than depend on the U.S. The British government was therefore only 'desirous of acquiring a share of the potential oil bearing lands throughout the world',[39] and added that there was no discrimination in the granting of concessions in British colonies. Moreover, American interests in Central and South America were in 'excess of the aggregate of those held by all other countries. So strongly entrenched is the United States in the great world reservoir of oil, that it would be well-nigh impossible to take the lead away from her'.[40] Consequently, any motives 'which have been ascribed to Great Britain of seeking to create a monopoly over the potential oil lands of the world are untenable'.[41] What was needed, Airey concluded, was 'competition of a healthy and stimulating character'[42] by all concerned. Nevertheless, while the convention was taking place Bainbridge Colby, U.S. Secretary of State, in a firm Note to Lord Curzon on 20 November wrote that

American oil companies had to be granted the right of developing the oil resources of Mesopotamia.[43]

On 6 January 1921, Senator MacKellar of Tennessee introduced a Bill in Congress to place an embargo on petroleum exported from the U.S. 'except to countries granting reciprocal treatment to the United States with regard to production and distribution of oil'.[44] Senator MacKellar said of Britain that:

> She, it seems, has plenty of money to invest in petroleum lands and to acquire fields throughout the world, but she does not have enough money to pay interest on the debts she owes the United States on account of great sums of money which had been loaned to her for the purpose of protecting her Empire . . . since the war the British Nation had acquired oil rights in Persia, in Mexico, in the United States itself, in Mesopotamia, in Palestine, in Venezuela, in Rumania [sic] . . . In other words, wherever in all the wide world she could acquire oil rights, the agents of her Government and her Nationals, with her aid and help, were attempting to acquire those oil rights.[45]

However,

> I admire her attempts to obtain control of the oil supply of the world. But I do not think we will be good Americans if we can stand idly by and let her gobble up the oil supply of the world if we can prevent it, and we can easily prevent it.[46]

Sir Auckland Geddes commented to Lord Curzon that the Senator's Congressional speech was a confused repetition of the statements 'put into circulation during the last twelve months by Standard Oil Company and the interests associated with them, relative to the alleged purpose of His Majesty's Government to secure for Great Britain a dominant position in the oil industry of the world'.[47] The Senator accused the British government and *Shell* of controlling all the oil deposits of Persia, of acquiring a majority shareholding in *Shell,* of preventing American oil companies from developing oil concessions secured in Mesopotamia and other countries, and of buying American oil at $1.75 to $2.25 per barrel and reselling it at $7.75 to $12.[48] On 17 January, Senator Phelan continued the attack when he introduced a similar Bill to place an embargo on American oil exported. These attacks had the clear objective of inducing foreign governments, especially Britain's, to give more favourable treatment to U.S. companies in foreign

13. Aerial view of Abadan Refinery – 1918.

14. Galicia, Russia. Derrick and engine house.

15. Rumania – crowded oil field 1927.

16. The Barroso blow-out in the La Rosa sector in Cabimas which occurred in December 1922. This is the beginning of the oil industry in Venezuela.

17. Curaçao c.1930. The harbour with the refinery in the background.

19. An early BP bulk fuel steam wagon – 1920.

18. Roadside filling station at Godalming (date unknown).

20. A Shell roadside pump in Clevedon, 1922.

21. Scammell Six Wheeled Articulated Heavy Fuel oil road tank
    wagon, Lansbury Wharf, Fulham c.1924.

22. Early BP road tanker – 1928.

oilfields from which they were excluded.[49] However, American exclusion from the limited number of oilfields in the Empire was exaggerated. Although there were no formal entry restrictions, foreign companies at present only exploited the oilbearing lands of British North Borneo, and the restrictions imposed by Canada and Trinidad only meant that foreign companies had to enter into partnership with British companies. If *Shell* had a large stake in Venezuela, American companies were more prominent in Columbia

> while in Peru, where the oil industry was at one time almost entirely British, the control of all companies but one has passed to the Standard Oil Company. Diplomatic influence was exerted by the United States to secure the exclusion of British interests from Colombia and Costa Rica and she can point to no similar action on the part of Great Britain to the detriment of American firms.[50]

In addition the British government felt that Venezuela's threat to rescind the concession held by the *Colon Development Co.* (a *Shell* subsidiary) for non-compliance of contract was inspired by *Exxon*. Sir J. Tilley of the Foreign Office in a Minute written on 11 May 1920, wrote that *Exxon*

> may be making determined efforts to turn out the British company in order to get the concession themselves. The policy to obtain concessions in foreign countries is one which has been followed by the Standard Oil Company for many years. They are at present irritated by our attitude in refusing them access during the period of occupation to the oilfields of Palestine and Mesopotamia, and there may presumably be reason for Standard Oil intrigues in South American States against British companies who happen to have obtained concessions in those countries.[51]

Most of Britain's oil requirements came from the U.S. which placed Britain in a serious position if the U.S. imposed an oil embargo. However, such a ban was unlikely, as by this time large stocks had accumulated owing to the post-war recession and to Mexican oil imports. As a result a total embargo at the time 'would probably add to the embarrassment of the industry and it seems on the whole unlikely that so extreme a course will be adopted'.[52] At this juncture the British government felt secure enough to tell the Americans that owing to their lion's share of the world's oil market it was not easy to justify the U.S. government's 'insistence that American control should now be extended to resources which may be developed in mandated territories, and that too at the expense of the subject of another State who have obtained a valid concession from the former Government of those territories'.[53]

Senator MacKellar found few supporters for his Oil Export Embargo Bill, 'while the occasional excursions into the area of Senator Phelan, of California, are attributed rather to his Irish ancestry and consequent bitter anti-British feeling than to serious convictions on the subject'.[54] Nevertheless, the anti-British attack in the U.S. was maintained in public. On 12 April 1921, Senator Lodge appealed to the U.S. government to grant as much protection to Americans as the British government did to her own citizens, and stated that:

> The indications are very strong that the very large oil fields, perhaps the largest in the world, are on the point of development in Venezuela and Colombia, but they will pass into the hands of the powerful British combination if our people cannot at least understand that they will be protected against wrong and injustice if they invest in countries other than their own for the purpose of furnishing the United States with oil and enlarging our commerce.[55]

Later Albert B. Fall, Interior Secretary, read out in Congress a letter sent from Senator Lodge which aserted that *Shell* was owned by the British government and controlled the 'oil fields of Venezuela which are developing by leaps and bounds'.[56]

Meanwhile, the oil men met in Britain to try and settle matters amongst themselves. At a meeting held at Stanbridge Earls, attended by Sir Charles Greenway, Chairman of *Anglo-Persian*, A.C. Bedford, President of *Exxon*, and Sir John Cadman of the Petroleum Department, Bedford expressed his desire to see British and American interests co-operating in commercial affairs and wanted to see the question of American participation in Mesopotamia removed from the diplomatic plane and onto a more commercial and practical one. Sir Charles agreed but pointed out that his company claimed a predominant position in Mesopotamia. The American acknowledged this but wanted to find a way of participating in *TPC*. A first step towards achieving this goal was made in June 1921 when the Inter-Departmental Committee on Petroleum agreed than an 'open door' policy should be followed.[57] Nevertheless, before this could take place the U.S. had to admit 'the validity of the Turkish Petroleum Company's claims'.[58] A further complication was that the British government had not reached a decision regarding who should develop the oilfields. Churchill, as Colonial Secretary, argued that there was no possibility for the British government to develop the oilfields directly.[59] In 1920, Sir R.S. Horne, President of the Board of Trade, considered that the British government did not possess the 'necessary organisation for so vast a business as the successful commercial exploitation of a large oilfield and the marketing of its products'.[60] As a result it would be far more satisfactory 'if the development were in the hands of a British company than in those of an Arab government over which our control will be hypothetical'.[61] The British government then decided that *TPC* should be reconstituted to exploit the region's oil sources. One outstanding stumbling block remained: the need to convince the U.S. of *TPC*'s legitimate claim to the oilfields of the area. The position held by the U.S. and based on the Chester claims and the heirs to Abdul Hamid, was that the relationship between *TPC* and Said Halim in the Vilayets of Mosul and Baghdad had been 'those of negotiators of an agreement in contemplation rather than those of parties to a contract'.[62] Consequently, the U.S. government did not recognise that *TPC* had ever held a concession. It wanted the matter settled by suitable arbitration.[63] These questions were discussed at the Washington Armaments Conference held from November 1921 to February 1922. The usual cries of 'open door' were made but no agreement was reached on *TPC*. It was, however, decided that the following points should be discussed at a future date: (a) oil conservation; (b) agreement among the Powers regarding the equal status of privilege to be granted nationals; and, (c) the right to explore for oil in all Colonies and Mandates.[64]

THE TURKISH PETROLEUM COMPANY

While negotiations were progressing and the dispute between the U.S. and the UK continued, the Near East witnessed a war between Turkey and Greece, the emergence of a new Turkey, and the repudiation of the Treaty of Sèvres. The question of the status of the Mosul Vilayet became pressing. These new political conditions brought back into the struggle the old Chester concession and delayed the solution to the Mesopotamian oil question for several years longer.

The British wanted the abolition of the Turkish Empire and the building up of new Islamic kingdoms around the Sherifian family with Hussein in Mecca, his youngest son Feisal in Damascus, and his older son Abdullah in Iraq. On the other hand French policy was aimed at restoring Turkey as a regional power 'and through her rebuilding French prestige in the Near East and among the Moslems as well as acquiring economic advantages in the new Turkey'.[65] On 20 October 1921, the French Foreign Minister, Franklin Bouillon, signed an agreement with his Turkish opposite, Musuf Kemal Bey, which provided Turkey with, *inter alia,* control of 400 miles of the Baghdad railway. The significance of this agreement was the recognition by France of the Turkish national movement and the termination of hostilities between the two countries. These events disturbed Britain and she accused the French of violating certain treaties against Turkey. At the end of the war the Greeks held on to Smyrna in Turkey and in 1922 under King Constantine they proposed to attack Constantinople. Allied support for this scheme did not materialise, and instead the Turks attacked and took Smyrna. This event also forced the drawing up of a new peace treaty between Turkey and the Allies.

On 22 June 1922, Bedford of *Exxon* with Acting Secretary of State Harrison discussed the Mesopotamian question. For Harrison, Mesopotamia was a testing-ground for America's 'open-door' policy, and consequently any arrangement 'which is not in agreement with that principle or which implies a repudiation of the view'[66] held by the U.S. government that the *TPC* 'has no valid concession in Mesopotamia, could not receive the approval'[67] of the government. Nevertheless, Harrison did not want to prevent American enterprise from 'availing itself of the very opportunities which our diplomatic representatives have striven to obtain'.[68] Thus he informed Bedford that the State Department would not object if British and American interests carried on private negotiations on the subject, provided that no American company which was willing to participate was excluded, and that *TPC*'s claim was validated by an impartial arbitration as previously suggested by the U.S. government. Bedford replied that all the American companies likely to be concerned with the Mesopotamian question were united behind him. The second point could be dealt with by awarding *TPC* a new confirmatory grant or concession. Soon after this, Bedford recommended to Sir Charles Greenway that negotiations be started between their respective companies on the following basis: (a) that as agreed by Allied Powers an 'open door' would be maintained in Mandated Territories; (b) that the State Department's contention regarding the invalidity of *TPC*'s claim would not be withdrawn but a new settlement be worked out using *TPC*'s concession as a basis for negotiations, culminating in the granting of a new concession covering the same area to be granted by the government ruling Mesopotamia; and, (c) American participation would be increased from the previous percentage offered.[69] Sir Charles agreed to these terms and *Exxon*'s representative was dispatched to London. In the months ahead the Cabinet met and decided that in the interest of all concerned parties the development of the oilfields would best be conducted privately. It was now a matter of persuading *Shell* and the French oil company which would jointly develop the concession, to agree to permit the entry of the U.S. The problem was that Deterding was 'more or less hostile to the Americans and he will not be anxious to allow them a participation merely to improve general relations with them or ease the diplomatic position'.[70] Nevertheless, he gave verbal assurances that the agreement being negotiated with the Americans would be acceptable, mainly because he was convinced that this would be the only way of making progress in Iraq. So French representatives arrived in London to discuss American participation as they were not opposed to it in principle. However, as M. Pineau of the French Delegation argued, some form of compensation would have to be secured for French interests in order for them to support American participation. Under Article 8 of the San Remo Agreement, the Iraqi interest could take up a maximum of 20 per cent of the shares of the company. If this option were exercised, then French interests would contribute half of the first 10 per cent, and the rest would be paid by the participants in proportion to their holdings. As the Agreement stood, if American participation were allowed the French would pay disproportinately more than the rest. It was feared that the French Parliamentary Commission set up to review the agreement would not view such disproportion sympathetically.[71] The British government agreed to modify the article provided a 'complete general agreement is reached in regard to American participation'.[72] The new protocol would permit American entry, while each group would contribute to Iraq's share in direct proportion to its own holdings.[73] Negotiations continued in order to resolve the final percentages of the share issue to be allotted to each participant. The proposed agreement was accepted by the State Department as being in the spirit of their 'open door' policy.[74] In addition, on 8 September 1922, the Board of Trade agreed that American interests should be allowed to participate in the development of Mesopotamia.[75] Nevertheless, the whole question of oil concessions in Iraq remained in suspense pending a decision in regard to the Mandate. When a settlement was reached on this point certain pre-war claims would have to be taken into consideration. Subject to these claims it would then be open to American and other interests to apply for concessions.

The Lausanne Conference, which began on 20 November 1922, was called to negotiate a peace treaty between Greece and Turkey as well as between Turkey and the Allies. For the interests of *TPC* the disturbing part in the Turkish–Allied relations was the uncertainty of the frontier between Turkey and Iraq, where the main oilbearing lands lay. Turkey claimed sovereignty over the provinces of Mosul, Suleimanieh, and Kirkuk. Ismet Pasha, the Turkish representative at the Conference, based the claim on the argument that the land belonged to the Kurds, who were a Turkish ethnographical, historical, and economic minority who sent representatives to the Ankara National Assembly. This placed *TPC* in a precarious position as it was doubtful that Turkey would recognise the company's Mosul concession. According to Shwadran, the first effect was for *TPC* to 'offer the American group a 24 per cent share if the United States supported the company's claim at Lausanne'.[76]

On 12 December 1922, after much discussion agreement was reached whereby American, French, *Shell* and *Anglo-Persian* were to hold 24 per cent each in *TPC*.[77] *Shell* accepted the agreement with a good deal of misgiving because it was subject to the proviso that *Exxon* secured the acceptance of the State Department 'as satisfying American claims to participation'[78] in Iraq, with the consequence that at the Lausanne Conference it would 'support strongly this arrangement to the exclusion of any other interests, American or otherwise'.[79] The State Department was unwilling to change its policy. In a statement Secretary Hughes notified Teagle that the Department did not recognize *TPC*'s claims, and that if a confirmatory grant were made it would not be automatically approved but would be examined on its own merits as any other concession. Hughes added that the Department could not take sides if the question of a concession's validity arose unless it was 'palpably without foundation'.[80] Hughes added:

> The effort of the Department is to maintain the Open Door and suitable opportunity for American enterprise. It is left to the American companies and individuals who are interested to take advantage of the opportunities that are offered and to promote their interests in any proper way. The Department is

always willing and desirous of giving proper diplomatic support to American interests, but if there are questions underlying the title and competing American claims you will readily understand that this Government cannot associate itself with one set of American claims as against another.[81]

The U.S. had never declared war on Turkey and therefore did not participate officially at the Conference. Where then would the U.S. stand if Mosul eventually fell into Turkish hands?

On 20 June 1922, Admiral Chester called on the State Department to seek support for his concession at the forthcoming Conference. It was generally understood at the State Department that Chester did not have legal claim. In 1922, however, the *Ottoman American Development Co.* was established once again and representatives sent to Turkey to negotiate a renewal of the Chester concession. But other American claims were involved. In 1918 the heirs to Abdul Hamid, 22 princes and princesses, signed an agreement with Captain Bennett, an American, by which all claims arising out of the personal property held by Abdul Hamid were to be turned over to a corporation organised to promote and develop them. Part of this property included the Mesopotamian oilfields. The Turkish National Assembly had made short shrift of these rights and the claims were not at all certain. Nevertheless, Captain Bennett together with Mr. Utermeyer, his lawyer, went to Lausanne to uphold the interests of Bennett's clients. As the Conference continued representatives of the Turkish delegation travelled to London to 'offer oil concessions in the Mosul area to British interests'.[82] The French opposed the adoption of the Chester concession by the Turkish government on the grounds that part of the railway line in the concession had been granted to French interests before the war.

British, American and Turkish interests in Mosul were not solely directed at the procurement of the oilbearing lands. Turkish policy was motivated by reasons of national prestige and military strategic considerations. The British were also motivated by strategic considerations of defence as Mosul and Iraq were considered as important links in the lifeline of Empire. American interests were prompted by principles of equality of opportunity for American nationals.

On 2 February 1923, the first Lausanne Conference ended with doubtful prospects of peace in the Middle East. The boundary question still remained unresolved and would be submitted to the

Council of the League of Nations for arbitration. In a last minute attempt to leave the Mosul Vilayet out of the settlement, Pasha requested it to be excluded from the Conference so that the matter could be settled by common agreement between Britain and Turkey.[83] The suggestion came too late as the Treaty had already been drafted, but Britain was willing to suspend its appeal to the League of Nations for one year provided that the *status quo* was preserved and there was no movement or alteration of armed forces, in order to allow the two countries to settle the matter amicably[84] and reach a settlement within nine months.[85] If such an agreement was achieved then the matter would not be submitted for arbitration to the League of Nations.

On 28 February 1923, Sir Henri Deterding, Sir John Cadman and Mr. H.E. Nicholls sailed on the ss *Majestic* for New York to enter into 'negotiations with the *Standard Oil Company* and its allied interests in arranging cooperation and plans for the future'.[86] By the end of April, good progress had been achieved and the question was 'removed from the political arena and dealt with on a purely business basis'.[87] Negotiations were moving on the basis of allowing more than one American interest into Iraq. Bedford explains:

> In the proper development of Mosul and Mesopotamia, American interests – not merely one American interest, but all important American interests – should cooperate with the interests of other nations, to the end that the risks involved in the investment of capital shall be widely distributed, and thus minimised; and also to the end that resources of that country shall not, any more than the resources of any underdeveloped section of the earth be exploited for the exclusive benefit of a single nation or group of interests.[88]

At the same time agreement between *Shell* and *Exxon* over Northern Persia was nearing completion. On 6 March 1916, Premier Sepaksalar granted to A.M. Khostaria, a Russian subject, a seventy-year concession to exploit oil in the five Northern Persian provinces. On 8 May 1920, *Anglo-Persian* bought it for £100,000 and organised the *Northern Persian Oil Co. Ltd.* with an authorised capital of £3 million. On 26 February 1921, Russia signed a treaty of alliance with Persia in which it *inter alia* denounced all treaties and conventions made between Tzarist Russia and Persia. The Persian government as a countermeasure to British and Russian influence tried to get the U.S. interested in exploiting the oil resources of the country. The Khostaria concession would be allowed to go to an American company as long as an American loan were available. On 22 November 1921, the Majlis abrogated the concession and voted unanimously to grant *Exxon* a fifty-year concession over the same region. Both Russia and Britain protested this vigorously: the Russians because the 1921 Russo-Persian Treaty (which had not been ratified) forbade the granting of a concession formerly held by a Russian citizen to a third party; and, in the British view, *Anglo-Persian's* concession was still valid since the Russo-Persian Treaty had not been ratified. *Anglo-Persian* too placed pressure on *Exxon*. In order for the American company to move their oil out a pipeline would have to pass through *Anglo-Persian's* concession over the rest of Persia which gave it exclusive right to transport oil. Without *Anglo-Persian's* 'participation Standard's' oil could not reach the world commercial markets'.[89] The British government contended that both companies would be best served by combining forces in Northern Persia, something which the two companies agreed to do in 1923.[90]

On the eve of the second session of the Lausanne Conference, held between 24 April and 24 July 1923, the Turkish government announced that it had approved the Chester concession. According to Shwadran, Secretary Hughes was delighted by this as it was a vindication of his 'open door' policy. The Turkish Ministry of Public Works and the *Ottoman American Development Co. Ltd.* concluded a contract for the company to rebuild Ankara as well as to construct certain railways, ports, and to have exclusive mining rights within a margin of twenty kilometres on each side of the railway lines. One of the proposed routes went through Mosul, Kirkuk and Suleimanieh, encroaching upon British Mandated Territory and threatening the Mosul oil rights.

Long and tedious discussions over *TPC's* rights, the French Silvas–Samsum railway and Vickers Armstrong's concession took place at Lausanne. The British government refused to admit that Mosul was part of Turkey and proposed to submit the whole question to arbitration at the League of Nations. Turkey for its part opposed such a scheme 'maintaining that she still had sovereignty over Mosul, and she would not submit her natural inheritance to arbitration'.[91] According to Shwadran, Lord Curzon threatened Turkey with war over her intransigent attitude, but the Conference ended with no reference to *TPC*. The thorny question of Mosul, however, was settled a few years later. When the Turks annulled the *Ottoman American Development Co. Ltd.*'s concession on 18

December 1923, because it did not attract sufficient capital, it became obvious that U.S. participation in the area would only be achieved through *TPC*. The French at the time urged the Turks to take a more conciliatory attitude over Mosul, which they did, agreeing to conduct negotiations between themselves and Britain over a period of one year in order to seek a solution to the problem. If within this period an agreement had not been reached then the question would be settled by arbitration at the League of Nations. On 19 May 1924, direct negotiations between Sir Percy Cox and Fethy Bey opened in Constantinople continuing until June without further progress, so that on 6 August 1924, the British government requested the League of Nations to decide the matter. As a result of border incidents and in order to preserve the *status quo* referred to in the Lausanne Treaty, the Council of the League of Nations decided to draw a provisional demarcation line. At the same time a Commissioner of the League was appointed to examine the vexing question. He recommended that the territory in dispute should remain an effective Mandate of the League for 25 years, or until in the opinion of the Council Iraq qualified as a Member of the League, and that due regard should be paid to the Kurdish nation's desires and aspirations. Both the Council of the League of Nations and Britain submitted to these recommendations, but Turkey questioned whether the Commission's recommendations were legally binding and appealed to the Permanent Court of International Justice at The Hague for a ruling. Once the Court declared on 21 November 1925, that the Council's decision was binding[92] Turkey felt it expedient to resign itself to Mosul's loss as by this time it was seeking membership of the League of Nations.[93] As a result of this the Council on 16 December 1925, decided that a new Turco-Iraqi frontier should be drawn,[94] and that a new treaty of Alliance between Iraq and the UK should be signed extending the UK's mandate for a further 25 years over the region,[95] something which was done on January 13 1926. On 6 June 1926, a further treaty was signed between the UK, Iraq and Turkey, in which the Mosul Vilayet was recognised by Turkey to form part of Iraq. The Americans endorsed this and stopped supporting American claims in the area.

In September 1923, a new accommodation between the oil companies and the Iraqi government was attained. *TPC* would be reorganised to allow equal participation by British, French, Dutch and American interests. In order to incorporate other interests wanting to participate in the region *TPC* would award a number of subleases. Such a system allowed the first group the best share of the oilbearing lands; moreover, the company was aware of the location of the best lands because of its extensive knowledge of the region. Despite this, Secretary of State Charles Hughes accepted the agreement as falling within the scope of his 'open door' policy, reasoning that 'while these other interests may not be in the first instance beneficiaries of the proposed concession they may, as leasees of the *concessionaire,* enjoy substantially equal rights in the development of Mesopotamian oil resources'.[96]

In order for the agreement to be effective it had to be endorsed by the Iraqi government. Under the San Remo Agreement the Iraqi government was entitled to 20 per cent of *TPC's* equity, something which had caused a great deal of friction amongst the four major partners. In 1922, the Iraqi government pressed its claim but *TPC* was willing to comply only if the government agreed to a substantial reduction in royalty payments. Although it supported the Iraqi government in principle, the Colonial Office held the same opinion as *TPC, viz* that royalty payments should be pegged at a much lower level, with the consequent result that the country would be no better off. The Iraqi government had pressed their claim because of the potentially high profits, but:

> The whole of the ordinary share of capital in the Turkish Petroleum Company was to be held by big oil corporations, and they had absolutely no interest in securing a particularly high rate of profit from their investment in the Turkish Petroleum Company. In fact, it would pay them perfectly well to earn no profit at all on such investment, if they obtain the oil from the company at a cheap rate.[97]

This was because American, British and French interests (representing 65 per cent of the total equity share) would be subject to British income tax, and therefore the smaller the profits the less tax paid.

A further difficulty was the division of the shares after Iraq's 20 per cent had been taken into account. The Iraqi government was adamant about their 20 per cent, threatening to disallow *TPC's* new concession if their percentage were denied. Despite this, the British government felt that Iraq had a clear international obligation to grant a concession to the company:

> The international position of the Company will result in several countries having a direct interest in the 'peace', order and good government of Iraq . . . It is, therefore, an inter-

national obligation of Iraq to grant a concession to the Turkish Petroleum Company, and it is very much in her interest that such a concession should be granted.[98]

The Cabinet decided to inform Iraq's High Commissioner that it was not the British government's intention to present an ultimatum or to press for the dismissal of the Iraqi government in order to overcome Iraq's objections to granting the oil concession, which would not provide for Iraq's 20 per cent claim. It was felt that negotiations should start with *TPC* to induce the company to increase its capital so as to take into account Iraq's 20 per cent share, but there were other possibilities open to the company. For instance, the Iraqi government could be given fully paid-up shares in *TPC* amounting to 20 per cent of the total issued share capital of the company. The shares would be of special ranking and given the same interest as ordinary shares, but the Iraqi government would not be permitted to intervene in the commercial affairs of the company. Any dividend paid on Iraq's shares would count against royalty payable to the government. The second possibility was for the whole question to be settled by arbitration, which the company preferred. On 7 March 1925, Sir Henry Dobbs, High Commissioner, was told to communicate *TPC*'s preference to the Iraqi government. As much pressure as possible was to be exerted on the Iraqi government to force their acceptance, but on the same day in a telegram to London, Sir Henry reported that the Iraqi government had arrived at a majority decision to instruct the Minister of Communications and Works to sign an agreement without insisting on the right of Iraq's participation in the share capital of the company. By 12 March, everything had been settled with the exception of one clause of the draft convention, one part relating to the price of oil for local consumption and the other giving the Iraqi government the option to acquire a refinery in certain circumstances.[99] After further negotiations *TPC* allowed this and a full concession was granted to a new subsidiary company, the *Iraq Petroleum Co.*,[100] on 14 March.

The aggressive role of the U.S. in securing participation for its nationals in *TPC* is disputed by various authors. Davenport and Cooke assert that Hughes 'won a diplomatic point over the British'.[101] Gibbs and Knowlton suggest that Britain allowed the U.S. such a stance because *inter alia* they needed American finance to rebuild Europe. Nash on the other hand states that the policies of Lansing, Bainbridge, Colby and Hughes 'revealed many weaknesses and vacillations',[102] and as a result 'American efforts to secure foreign oil concessions in the Middle East and Sumatra met only minimal success'.[103] Nevertheless, it is doubtful if the U.S. would have achieved any participation at all without Hughes' persistence. It was now clear that from henceforth the U.S. would have to be taken into account during the formulation of any major international policies which might clash with American interests. The U.S. insisted on its right to an 'open door' policy not because she was threatened with an oil shortage (a justification used forcibly in the early stages, but later losing weight due to the enormous oil glut building up in the U.S.), as she controlled 80 per cent of the world's oil production, but in order to ensure that her citizens were given equal economic participation in the development of the former Turkish Empire. The underlying objective, made absolutely clear, was that the U.S. would not accept the prohibition of her citizens' participation in the economic development of areas hitherto barred to American capital. The large integrated oil companies, too, learnt a lesson which was that co-operation (with their respective government's backing) would achieve greater profits for their shareholders against the smaller oil companies. This co-operation was to be further consolidated during the ensuing years of the 1920's. For the British government, however, the outlook was dull. Its policy to lessen its dependence on American oil through a majority British stake in *Shell* and the encouragement of Middle Eastern oil production did not live up to expectations. For the rest of the decade oil production from the Middle East was further delayed by international disputes and political instability.

THE IRAQ PETROLEUM CO.

The 1925 Convention between *IPC* and the Iraqi government stipulated that within 32 months the company would select 24 blocks of eight square miles each for the exclusive exploitation of oil. When the company struck oil it decided to concentrate on oil production rather than continuing prospecting. The construction of a pipeline to the Mediterranean was an essential element in these plans, but before this could be built the company needed to free itself from the 1925 Convention by which it agreed to select its exploitation blocks up to November 1928 before it could start development work. On 11 November 1927, Sir Adam Ritchie, *IPC*'s General Manager, advised Sir Henry Dobbs that his company was contemplating building a pipeline to the Mediterranean and needed the latter's assistance to get the 1925 Convention rescinded. The Foreign Office was not in favour as the company would have to have its concession extended. This would involve a postponement of the date by which the government might offer the rejected *IPC* blocks in open competition. It would also give rise to criticism from the U.S. Nevertheless, on 8 March 1928, IPC asked formally for a five year extension to its prospecting period,[104] something which the Iraqi Cabinet approved. But such an extension would only be effective after the King's approval and ratification by the Iraqi Parliament had been secured. At this stage, the *British Oil Development Co.*[105] offered to construct a railway from Baghdad to Haifa without any financial subsidy from the Iraqi Government if given an oil concession. The group was formed by Lord Wester Wemyss, Admiral of the Fleet, and Colonel Stanley, who were keen to acquire *IPC*'s concession. If this were not achieved they would 'oppose by all means in their power the grant by the Iraq government'[106] of *IPC*'s five-year extension because once the extension had been granted then all other interests would be excluded from the region for a further five years. *BOD*'s offer to construct a railway was of greater interest to King Faisal than the oil pipeline because it would contribute more to the prosperity and economic development of the country. In May, Lord Wemyss travelled to Baghdad to set his plans for *BOD*'s railway before the King and his government. At the same time, *IPC* made counter-offers and took a renewed interest in the railway project. On 6 July 1928, Sir Adam Ritchie saw King Faisal and the Prime Minister, and as a result of the meeting informed his company that

> such strong feeling existed in Baghdad about the necessity for a railway to Haifa that he recommended that, in return for their five-year extension, a definite assurance should be given by the Turkish Petroleum Company that, if they constructed a pipe-line to the Mediterranean, they would by some means simultaneously arrange for the construction of a railway.[107]

On 18 July the Iraqi government rejected the latter part of *IPC*'s offer as being too indefinite, but the company did achieve a minor extension of a few months. Almost simultaneously, on 31 July 1928, the Red Line Area Agreement was signed by the companies which formed *IPC*. The agreement stipulated that individual members of *TPC* could only operate or acquire concessions in the former Ottoman Empire through *TPC*. The area covered by the agreement extended over Iraq, Turkey and the Saudi Arabian peninsula (excluding Kuwait), an area which was to become a major oil producing province.

American participation in *IPC* was secured by the formation of the *Near Eastern Development Corp.* in 1928. Its shares were divided among the following American companies: *The Atlantic Refining Co.* (16.67 per cent); *Gulf* (16.67 per cent); *Pan American Petroleum & Transport Co.* (16.67 per cent); *Exxon* (25 per cent), and *Standard Oil Co. New York* (25 per cent).[108] This company held 23.75 per cent of the *TPC* shares, while *Shell*, *Anglo-Persian* and French interests each took 23.75 per cent of the shares, with Calouste Gulbenkian securing 5 per cent. The members of *TPC* agreed to operate within the Red Line Area only but American companies could bid for subleases in territory rejected by *IPC* in Iraq. In areas outside Iraq

> but still within the Red Line area, the American signatories might bid for concessions; if, however, they were successful in their bids they were not to engage in operations on the concessions obtained until they first offered the other members of the Turkish Petroleum Company the right of equal participation. In so far as any oil company not a party to the Red Line Agreement was concerned, the Open Door concept as it applied to Iraq was accorded nebulous recognition at best, though the agreement could not prevent any such company from engaging in activities outside Iraq. In return for their interest in the Turkish Petroleum Company and its promising concession in Iraq, the five participating American companies in effect renounced their claim to independence of action anywhere within the boundaries of the old Turkish Empire.[109]

In January 1930 a tripartite convention was signed between the U.S., Britain and Iraq in which the U.S. recognised Iraq, and its nationals were guaranteed all rights and benefits secured to members of the League of Nations.[110] However, agreement between the company and Iraq was hindered by the international character of the company. From a defence point of view the British government wanted to see the railway and pipeline terminate at Haifa. For its own part, the Iraq government also had strong reasons for wishing the railway to end at Haifa because it meant that the railway would have to run through Trans-Jordan, an area ruled by King Faisal's brother, thus bringing economic benefits to the country. It would also provide defence against Ibn Saud by preventing his northward expansion which would cut Iraq off from

Trans-Jordan.[111] On the other hand French interests in *IPC* opposed the Haifa terminus, preferring one at Tripoli in Syria, a French Mandate.

On 24 March 1931, an agreement was signed giving the *IPC* concession more liberal terms, increasing it from 192 square miles to 35,126 square miles. The company was freed from drilling obligations and thus could devote all its time to exploration.[112] In June of the same year, the Iraqi government put all other lands up for bids, granting on 20 April 1932, a 75-year concession to *BOD* covering 41,302 square miles.[113] However, by 1937 *IPC* had effective control of *BOD* and, in 1941, the *Mosul Petroleum Co.* took over *BOD*'s concession. In 1938, the *Basrah Petroleum Co.*, a subsidiary of *IPC*, obtained a concession covering the remainder of Iraq, that is, some 87,236 square miles and contiguous to Kuwait. By 1938, then, *IPC* had control over the entire country except for the small district of Khanagin which was covered by the 1901 D'Arcy concession. Nevertheless, extensive exploitation of the oilfields was delayed until after the Second World War

PERSIA

In 1927, the Persian government demanded not only the invalidation of the Armitage-Smith agreement, but also the D'Arcy concession. Sir John Cadman wrote to Prince Teymourtache, Persian Minister of Court, that an extension of the concession period would be necessary if the requisite capital were to be obtained. In his reply, the Prince stated that a new contract had to be negotiated as the initial concession had been granted when the full implications of oil were unknown to the Persian government.[114] Further discussions took place and in 1932 Sir John Cadman informed the Persian government that a revision of the concession could not be contemplated. As the company had persistently refused to recognise Persia's rightful complaint there was no other alternative but to cancel its concession. The Persian government was still prepared to enter into negotiations for a new concession, but *Anglo-Persian* vehemently denied the Persian government's allegations, and the British government protested vigorously against this action. The British preferred that the matter be settled by arbitration at the League of Nations. Nevertheless, on 27 November 1932, the concession was rescinded by the Persian government.

Early in 1933, Sir John Cadman arrived in Teheran to negotiate a new agreement. At first his terms were unacceptable, but on 28 April

> Cadman and the British Minister had an audience with the Shah. It was during this interview that the Shah was forced to surrender. He was told very frankly that his refusal to consent to the terms of the British proposals would bring about the rupture of relations between the two countries'.[115]

Reza Khan, the Shah, was not prepared to plunge the country into chaos as he knew that the British would occupy Khuzestan and this might provoke the Russians to take over Azerbaijan. Consequently, on 1 May 1933, a new agreement was announced which reduced the company's concession from 450,000 square miles to 100,000 square miles. The company's pipeline monopoly was abolished and the government was guaranteed a minimum royalty payment of £1 million per annum. *Anglo-Persian* would now pay four shillings per ton royalty instead of 16 per cent on profits. This, together with the minimum royalty clause, was a considerable advance to the government in times of world-wide economic depression. The government were, however, entitled to 20 per cent of the company's dividends over £671,250. Also the concession was increased by 30 years. Finally, the government undertook not to cancel the concession nor alter its provision by legislative or administrative measures in the future. The company did not object to the reduction of its land because by 1932 it knew well enough where the best oilbearing land was to be found.[116] The agreement also had an added advantage:

> It disposed of a source of constant criticism of the Iranian Government that the company had a monopoly of oil over a large area and did not attempt to develop fully its possible resources. Further, the company gained good-will for its apparent liberality which might also prove a good bargaining point.[117]

The new concession did not contain a time limit for exploration, nor a surface tax on undeveloped parts to encourage company exploration. The period which followed this agreement saw an improvement in Anglo-Persian relations.

The high hopes entertained by Britain that the Middle East would become a large oil-producing region and so relieve her dependence on American oil imports did not materialise. The intricacies of regional politics and the problems of logistics associated with the region prevented it becoming the major oil producer the oil industry as a whole expected. It was Venezuela that was to develop during the 1920's as a major oil producer providing Britain with an alternative to her traditional sources of supplies.

NOTES

1. Deleted.
2. FO 371/4585 Des. 989 Geddes (A.) to Lord Curzon, 20.7.20. The scaremongers did not distinguish too closely *Shell*'s parentage. For the Americans *Shell* was a British company.
3. *Ibid.*
4. U.S. House of Representatives, *Petroleum Investigation,* 1934, Statement by Arthur H. Redfield (Bureau of Mines, Washington).
5. *Ibid.,* p.151.
6. *Ibid.,* p.151. On 16 August 1919 the State Department sent a Circular to all its Consuls stating that:
   > The vital importance of securing adequate supplies of mineral oil both for the present and future needs of the United States has been forcibly brought to the attention of the Department. The development of proven oil fields and exploration of new areas is being aggressively conducted in many parts of the world by natives of various countries and concessions for mineral oil rights are being actively sought. It is desired to have the most complete and recent information regarding such activities either by United States citizens or by others.
   and
   > Care should be taken, however, to distinguish between United States citizens representing United States capital and United States citizens representing foreign capital; also between companies incorporated in the United States and actually controlled by United States capital and those companies which are merely incorporated under United States Laws, but dominated by foreign capital. FO 371/5641.
7. U.S. House of Representatives, *Petroleum Investigation,* 1934, Part 3, Statement by Wirt Franklin.
8. U.S. Federal Trade Commission, 'Advance in the price of petroleum products', *House Report No. 801,* 66 Cong. 2 Ses. (Washington, 1920). Mexico supplied 99 per cent of U.S. imports at the time, accounting for 60 per cent of Mexico's total production (*ibid*).
9. FO 371/4585 Des. 909 Geddes to Ld. Curzon, Washington, 29.7.20.
10. *Exxon* companies owned nearly 37 per cent of the refineries, the 16 largest Independents owned a further 32 per cent, and the smaller refiners controlled the remaining 31 per cent of the market (cf. Federal Trade Commission, 'Advance in the price of petroleum products,' 1920).
11. 'Giant Struggle for Control of World's Oil Supplies', *New York Times,* 27.6.20.
12. U.S. House of Representatives, *Petroleum Investigation,* 1934 Part 3, p.151.
13. 'Oil a casus belli', *Journal of Commerce* (New York), 6.8.26.
14. The essential ingredients of the 'open door' policy were: first, that the nationals of all countries, in Mandated Territories, be subject to equal treatment in law; second, that no economic concessions in any Mandated region be so large as to

be exclusive; and, third, that no monopolistic concession relating to any commodity be granted.

15. Cf. Nash, *The United States Oil Policy*.
16. In January 1921 *Royal Dutch* acquired all the shares of the *Shell Co. of California*, which it then merged with the *Union Oil Co.* of Delaware (72 per cent of the shares going to the *Shell Group*, and 28 per cent to *Union Oil*) in order to avoid the operation of the Leasing Act. (FO 371/5641 G. Haly to Mr. Weakley, 11.11.21.) The U.S. felt strongly that Britain had barred American companies from exploiting the oil resources of the British Crown lands. In 1902 the *Colonial Oil Co.* of New Jersey (an *Exxon* subsidiary) applied for a licence to prospect for oil in Burma, which was refused. The same year the *Anglo-American Oil Co.* (a British registered *Exxon* subsidiary) was refused permission to prospect and exploit oil in India. In 1917 *Exxon* sought to buy the oil rights from private concessionaire holders in Assan, India, but the British government withheld it. (cf. U.S. Federal Trade Commission, *The International Petroleum Cartel, op. cit.*)
17. POWE 3/353 H.G. Chilton to G.R. Warner, 29.6.23
18. POWE 33/353 Des.371, Geddes (A.) to Lord Curzon, 23.3.23.
19. FO 371/5684 Des.134, Geddes (A.) to Lord Curzon, 2.2.21.
20. FO 371/4587 Des.1382, Geddes (A.) to Lord Curzon, 26.11.20, Encl., *Washington Post*, 18.11.20.
21. L.C. Snyder, 'The Petroleum Resources of the United States', *Proceedings of the Academy of Political Sciences*, 12:1 (July 1926), 159-167, p.166.
22. Cf. Francisco Alonzo González, *Historia y Petróleo, México: El Problema del Petróleo* (México: Editorial Ayuso, 1972); Lorenzo Meyer, *Mexico y Estados Unidos en el conflicto petrolero, 1917-1942* (México; El Colegio de México, 1968); and, Merril Rippy, *Oil and the Mexican Revolution* (Leiden: E.J. Brill, 1972).
23. Venezuelan crude first entered the U.S. in large quantities in 1926 when 12.5 million barrels were imported, rising to 50.6 million barrels in 1929. In 1925 Venezuelan crude accounted for nearly 2 per cent of total U.S. crude oil imports, by 1936 it accounted for 70 per cent.
24. Nash, *op.cit.*
25. 'Correspondence between His Majesty's Government and the United States Ambassador respecting economic rights in Mandated Territories' (Misc. No.10), *PP* 1921, Vol.xliii, Cmd. 1226, pp.481-94, John W. Davis to Lord Curzon, 12.5.20.
26. *Ibid.,* Lord Curzon to Davis, 9.8.20.
27. *Ibid.*
28. *Ibid.*
29. *Ibid.* Lord Curzon gives as examples the occupation by U.S. Marines of Haiti in 1913, when the U.S. refused to confirm oil concessions granted by the previous Haitian government to a British subject. In Costa Rica the U.S. Minister urged the San José government that succeeded President González Flores to cancel the concession granted by that government to the *British Controlled Oilfields Ltd.*
30. Davis to Lord Curzon, 12.5.20 in 'Correspondence between H.M. Government and the U.S. Ambassador' *op.cit.*
31. *Ibid.*
32. *Ibid.*
33. *Ibid.,* Davis to Lord Curzon, 28.7.20.
34. FO 371/4585, *San Francisco Bulletin*, 23.7.20. Enclosed in Des. 1089 Geddes (A.) to Lord Curzon, 10.8.20.
35. FO 371/4587 Telegram, Geddes (A.) to Lord Curzon, 21.11.20. Sir Auckland

felt that the Convention was organised by men who 'are amongst the strongest financial supporters of Sinn Fein', and had *Exxon* backing.

36. FO 371/4587 *Washington Post*, 18.11.20. Enclosed in Des.1382 Geddes (A.) to Lord Curzon, 26.11.20.
37. *Ibid.*
38. FO 371/4587 'Petroleum needs a World Question', *Journal of Commerce & Commercial Bulletin*, 18.11.20.
39. FO 371/4587 *American Petroleum Institute Bulletin*, No. 125, 18.11.20.
40. *Ibid.*
41. *Ibid.*
42. *Ibid.*
43. Bainbridge Colby to Lord Curzon, 20.11.20, in 'Correspondence between His Majesty's Government and the U.S. Ambassador'. The feeling that *Exxon* would enter the Mesopotamian region by fair or foul means was reinforced at the time when the *Standard Franco Americaine Co.* was formed in France with a capital of FFr. 10 million, divided equally among French and American shareholders. The French capital came from the *Banque de Paris et des Pays Bas* and a certain refining company. The American capital was believed to have originated from *Exxon* sources, and it was rumoured that *Exxon* through this company would secure a portion of France's 25 per cent stake in *TPC*. H.J. Seymour, Britain's Ambassador in Paris, reported that this was 'somewhat doubtful'. (POWE 33/95 H.J. Seymour to Lord Hardinge, Paris, 12.1.21; and, 'Arrival of American Oil in Europe and Colby's Note Produces Effect', *Oil, Paint and Drug Reporter*, 20.12.1920.
44. FO 371/5638 Teleg. Geddes (A.) to Lord Curzon, 10.1.21.
45. *Congressional Record*, 66 Cong., 3 Ses., 6.1.21.
46. *Ibid.*
47. FO 371/5638 Des. 66 Geddes (A.) to Lord Curzon, 19.1.21.
48. *Ibid.*
49. FO 371/5684 Des. 134 Geddes (A.) to Lord Curzon, 2.2.21.
50. FO 199/229 Petroleum Dept., 'Memorandum on the Petroleum Situation', 10.2.21.
51. FO 371/4623 Minute written on Telegram C. Dormer to Foreign Office, 4.5.20.
52. FO 199/229 Petroleum Dept., 'Memorandum on the Petroleum Situation', 10.2.21.
53. Lord Curzon to Davis, 28.2.21, in 'Correspondence between H.M. Govt. and the U.S. Ambassador'.
54. POWE 33/176 Teleg. 139 Conf. Geddes (A.) to Lord Curzon, 11.3.21.
55. *Congressional Record*, Vol. 61, Pt. 1, 67 Cong., 1 Ses., p.162. The previous year Senator Lodge appealed on behalf of the *General Asphalt Co.* for State Department assistance for the *Colon Development Co.*, a *Shell* subsidiary, in the dispute it was having with the Venezuelan government. (cf. B.S. McBeth, *Royal Dutch-Shell vs. Venezuela*,) (Oxford: Oxford Microform, 1982).
56. FO 371/5639 Teleg. Geddes (A.) to Foreign Office, 14.4.21.
57. FO 371/13540 H.W. Cole (Pet. Dept) to R.L. Craigie, 8.7.29.
58. POWE 33/95 Cadman, 'Memorandum of a meeting held at Stanbridge Earls on April 9, 1921, at which Sir Charles Greenway, Mr. A.C. Bedford, and Sir John Cadman were present'.
59. CAB 24/125 W. Churchill to Hankey (Secretary of the Cabinet), 20.6.21. Churchill felt that only the government of Mesopotamia could do this.
60. POWE 33/176 Sir R.S. Horne, 'Cabinet. Mesopotamian Oilfields. Memorandum by the President of the Board of Trade', 16.4.20.
61. CAB 24/125 Churchill to Hankey, 20.6.21.

62. Des. 287 George Harvey (U.S. Ambassador) to Lord Curzon, 17.11.21, in U.S. Senate, 'Oil Concessions in Foreign Countries', *Senate Doc. 97*, 68 Cong., 1 Ses., 23.4.23, p.50.
63. *Ibid.*, 'Memorandum of American Embassy (London) to British Foreign Office, August 24, 1921, entitled "Position of the Government of the United States concerning mandates".'
64. POWE 33/93 Sir Philip Lloyd-Greame, 'Notes', Undated; and, 'Conference to hear Oil Plans', *Oil, Paint and Drug Reporter*, 9.11.21.
65. Shwadran, *op. cit.*, p.220.
66. Teleg. 185 Acting Secretary of State (Harrison) to Ambassador in Great Britain (G. Harvey), 24.6.22, in U.S. Department of State *Papers relating to the Foreign Relations of the United States*, (1922), Vol. 2.
67. *Ibid.*
68. *Ibid.*
69. *Ibid.* The Chairman of the Board of Directors, *Exxon* (A.C. Bedford) to the Chairman of the *Anglo-Persian* (Sir Charles Greenway), New York, 26.6.22.
70. POWE 33/95 Clarke to Sir John Schuckburgh (Colonial Office), 5.7.22.
71. POWE 33/95 'Reopening of Discussions which took place Summer of 1922', Undated.
72. POWE 33/95 Geoffrey Hay to Major Lautier (French Embassy Official), 29.7.22.
73. POWE 33/95 Clarke to Secretary of State (Colonial Office), 28.7.22.
74. Charles E. Hughes to Teagle, 22.8.22, in *Foreign Relations* (1922), *op. cit.*
75. POWE 34/1 Board of Trade, 'Notes of a Conference held in the President's room with reference to oil concessions in Mesopotamia', 8.9.22.
76. Shwadran, *op. cit.*, p.220.
77. Calouste Gulbenkian received 4 per cent (later increased to 5 per cent) compensation for a concession acquired from the Ottoman government before the war.
78. Montague Piesse (London representative of *Exxon*) to Teagle, in *Foreign Relations*, 1922.
79. *Ibid.*
80. Hughes to Teagle, 30.12.22, in *Foreign Relations*, *op. cit.*
81. *Ibid.*
82. Davenport & Cooke, *op. cit.*, p.147.
83. 'Lausanne Conference on Near Eastern Affairs, 1922–1923. Records of Proceedings and Draft of Terms of Peace (Turkey) No. 1', *PP*, 1923, Vol. 26, Cmd. 1814, pp. 1-861. Ismet Pasha to the Presidents of the British, French and Italian Delegations, Lausanne, 4.2.23, 'Memorandum', pp.837-41.
84. *Ibid.*, No.56 British Secretary's Notes of Meeting, held in Lord Curzon's rooms at the Beau Rivage Hotel, Lausanne, 4.2.23, pp. 842-53.
85. 'Treaty of Peace with Turkey, and other instruments signed at Lausanne on 24 July 1923, together with agreements between Greece and Turkey signed on 30 January 1923, and subsidiary documents forming part of the Turkish Peace Settlement' (Treaty Series No. 16 (1923)), *PP* 1923, Vol. 25, Cmd. 1929, pp. 533-784.
86. FO 371/8593 Conf., G. Armstrong to Geddes (A.), 15.2.23.
87. FO 371/8494 Des.495 Geddes (A.) to Foreign Office, 20.4.23.
88. A.C. Bedford, 'The World Oil Situation', *Foreign Affairs*, 1:3 (15 March 1923), 96-107, p.104.
89. Shwadran, *op. cit.*
90. The Persian government after a long delay rejected the joint exploration offer, and on 20 December 1923, granted Sinclair a concession after he had promised to obtain a $10 million loan for the Persian government. *Anglo-Persian* refused

to grant Sinclair permission to transport his oil over its concession. Sinclair would, therefore, have to transport the oil over Russian territory. There were no obstacles to this as Sinclair was on good terms with the Russians, but soon after the Teapot Dome scandal in the U.S. erupted where it was shown that Sinclair had bribed certain American officials, Russian support for his scheme was withdrawn. The result was that Sinclair was unable to exploit his Persian concession.

91. Shwadran, *op. cit.*, p.226.
92. 'League of Nations. Report by M. Unden on the question of the Turco-Irak frontier, Geneva, 16 December 1925 (Misc. 16, 1925)' *PP*, 1924-5, Vol. 31, Cmd. 2565, pp.563-72.
93. Shwadran, *op. cit.*, p.238.
94. 'League of Nations. Decision relating to the Turco-Iraq frontier adopted by the Council of the League of Nations, Geneva, 16 December 1925 (Misc. 17 (1925))', *PP*, 1924-5, Vol. 31, Cmd. 2562, pp.549-54.
95. 'League of Nations. Thirty-Ninth Session of the Council. Report by the Rt. Hon. Sir Austen Chamberlain K.G. M.P. Misc. 4 (1926))' *PP* 1926, Vol. 30, Cmd.2646, pp.637-42. In 1932 Iraq ceased to be a British mandate. (cf. 'Protocol between the Government of the United Kingdom, France and Iraq for the transfer from the United Kingdom to Iraq of certain rights and obligations under the San Remo oil agreement of 24 April 1920, and the Convention between the United Kingdom and France of 23 December 1920 relating to Mandates in the Middle East', Geneva, 10 October 1932 (Treaty Series No. 37 (1932)', *PP*, 1932-3, Vol. 27, Cmd. 4220, pp.401-2).
96. Hughes to Teagle, 8.11.23, in *Foreign Relations*, (1923), Vol. 2, 'Great Britain', pp.218-318.
97. CAB 24/172/C.P.108 (25), Cabinet. L.S.A., 'Iraq: Turkish Petroleum Company. Memorandum by the Secretary of State for the Colonies', 23.2.25.
98. *Ibid.*
99. CAB 24/172/Conf./C.P.171 (25) L.S. Amery, 'Cabinet. Iraq: Turkish Petroleum Company. Memorandum by the Secretary of State for the Colonies', 18.3.25.
100. Hereinafter *IPC*.
101. Davenport & Cooke, *op. cit.*, p.155.
102. Nash, *op. cit.*, p.49.
103. *Ibid.*
104. CAB 24/202/C.P. 80(29) Secret, 'Cabinet. Oil Position in Iraq. Historical Note'. Undated.
105. Hereinafter *BOD*.
106. CAB 24/195/C.P.164 (28) Cabinet, 'Conflicting oil interests in Iraq. Memorandum by the Secretary of State for the Colonies. Initialled L.S.A.', 22.5.28.
107. CAB 24/202/C.P.80 (29) Secret, 'Historical Note . . .', *op. cit.*
108. Gibb & Knowlton, *op. cit.*, p.306.
109. *Ibid.*, p.306-8.
110. cf. U.S. Senate, 'Diplomatic protection of American Petroleum interests in Mesopotamia, Netherlands East Indies, and Mexico', *Senate Doc. No. 43,* 79 Cong., 1 Ses., 1945; and, 'Convention between His Majesty and His Majesty the King of Iraq and the President of the United States of America regarding the rights of the United States and of its nationals in Iraq, with Protocol and Exchange of Notes, London, 9 January 1930' *PP*, 1930-31, Vol. 34, Cmd. 3833, pp.23-76.
111. The other reasons given by the Iraqi government for wanting the terminal at Haifa were that it offered the longest haul of rail and length of pipeline in Iraq,

which was preferable from an economic and political standpoint; Iraq, Trans-Jordan and Palestine were all connected to Britain, whereas Syria was associated with France. It was also felt that it was better to be dependent on one Power than on two; Iraq wanted to develop trade with Egypt; southern alignment was more valuable to Iraq; and public opinion opposed the Syrian route.

112. According to Mikdashi *BOD*'s concession was more favourable to Iraq, but did not get off the ground, and illustrates the thin line between success and failure for a government trying to increase its revenue.

113. Royalties to Iraq were 4 shillings gold per tonne produced, with the requirement of producing £400,000 (gold) for the first 20 years beginning with the first cargo of oil shipped. Until this occurred the company would pay £400,000 (gold) per annum, half of which would be deducted from future royalty payments. The company was exempted from taxation in consideration of a yearly payment of £9,000 (gold) up to the time of commercial exports and thereafter £60,000 for the first 4 million tonnes produced, and £20,000 (gold) on each additional million tonnes produced.

114. According to Isawi and Yegaweh, the Persian government received between 1926–9 on average 12 cents per barrel. (Charles Isawi & Mohammed Yeganeh, *The Economics of Middle Eastern Oil* (London: Faber & Faber, 1963), p.128.

115. Nasrollah Saipour Fatemi, *Oil Diplomacy. Powderkeg in Iran* (New York: Whittier Books Inc., 1954), p.180.

116. Sir John Cadman at *Anglo-Persian*'s 24th. Annual Meeting held on 22 July 1933, stated that their new 100,000 square miles concession were enough for the company.

117. Nikpay, *op.cit.*, p.517.

4

Venezuela: Britain's New Oil Supplier

As we have seen, during the initial years after the Great War the industrialised world faced what appeared to be a real oil shortage and looked accordingly for a cheap and geographically convenient source of oil. At first it was felt that the Middle East would become a major oil supplier, but political and logistic problems delayed development during the 1920s. It was necessary to find a new major supplier free from the political and technical problems associated with Mexico. Other Latin American countries such as Colombia, Peru and Argentina appeared to offer promising prospects. For example, in 1916 the *Tropical Oil Co.*, an American concern, acquired in Colombia the De Mares oil concession; later another American company, the *Colombian Oil Corp.*, acquired the oil concession awarded to General Virgilio Barco. Both companies later entered into a development agreement with the *Gulf Oil Corp.* of Pittsburgh. Lord Cowdray, head of *S. Pearson & Sons Ltd.*, with strong Mexican oil interests, was also interested in developing Colombia's oil resources. However, the Executive Decree of 20 June 1919 on oil alarmed the oil companies because, in the opinion of the U.S Senate, if strictly enforced it would amount to confiscation of the oil companies' property.[1] Similarly, in Peru, American companies had shown an early interest in the oil deposits of the country. In 1913 *Exxon* purchased the British company of *London & Pacific* whose oil operations were centred in La Brea and Pariñas Estate. A new company, the *Imperial Oil Co.*, registered in Canada was formed to exploit these oil deposits. However, the company ran into trouble with the government over higher taxes on production, which was only resolved by arbitration in 1922.[2] In the same year *Exxon* acquired from the *Richmond Levering Co.* a one million hectares concession in Bolivia.[3] The government of Argentina, fearful that foreign oil companies would exploit their oil, was quick to enact legislation preventing this.[4] In the case of Brazil, the oil companies encountered difficulties with the government as the latter did not wish to encourage foreign oil company development of its potential oil sources.[5]

On the other hand Venezuela during the decade after 1910 experienced a minor boom in oil development, and by the end of World War I there were great expectations among government officials that the country would become a major oil producer. The initial entry of *Shell* during this decade ensured that the country's oil resources would be developed, and by the early 1920s the Group was poised for a massive increase in oil production. The large American companies such as *Exxon, Gulf Oil* and *Standard Oil Co. (Indiana)* soon followed suit, removing all doubts that existed as to whether the country would become a major oil producer. As we can see from Table II below, production took off in 1925, and within the short period of three years the country had become the second largest oil producer and first oil exporter of the world.[6]

TABLE II

Venezuelan oil production, 1917–36 (in 000s barrels)

Year	Production	% increase over the preceding year	% of total world oil production
1917	121	–	0.02
1918	321	5.3	0.06
1919	305	–4.9	0.06
1920	462	51.5	0.07
1921	1,449	213.6	0.19
1922	2,235	54.2	0.26
1923	4,327	93.6	0.43
1924	9,129	110.9	0.90
1925	19,933	118.4	1.86
1926	35,654	78.9	3.25
1927	60,419	69.5	4.79
1928	105,957	75.4	8.00
1929	136,074	28.4	9.16
1930	135,246	–0.6	9.59
1931	116,873	–13.6	8.52
1932	116,737	–0.1	8.91
1933	118,199	1.3	8.19
1934	136,287	15.3	8.95
1935	148,516	8.9	8.98
1936	154,639	4.1	8.63

Source: Adapted, Venezuela, Ministerio de Minas e Hidrocarburos, Dirección General, División de Economía Petrolera, *Petróleo y otros datos estadísticos* (Caracas, 1964), pp.138-9.

BOP–D*

By 1923, Britain's consumption of petrol, fuel and gas oil had increased enormously since pre-war days. Oil was to be an important energy source from that time on. Extensive refineries were established in Britain to cope with demand.[7] By 1925 dependence on the U.S. for crude oil and fuel oil, but not for petrol, appeared to be diminishing, as Table III shows.

TABLE III

UK's oil imports, 1925 (per cent)

Country	Crude oil	Petrol	Fuel oil
Persia	43	16	–
Curaçao (Venezuela)	20	–	–
Mexico	14	–	49
USA	–.	47	24
Soviet Union	–	7	10

Source: U.S. Department of Trade, 'British Petroleum Trade in 1925', Bureau of Foreign & Domestic Commerce, *Trade Information Bulletin No. 407*, April 1926.

The U.S. was still the largest producer of oil in the world. As we saw, the Americans campaigned hard in the immediate post-war period for the inclusion of American companies in the exploitation of the oil resources of the British colonies. In order to pressure the British government to recognise this, they enacted a number of laws, the most famous being the 1920 Mineral Oil Leasing Act. This legislation placed *Shell*'s and Lord Cowdray's interests in the U.S. at risk. Sir Auckland Geddes reported from Washington that, due to the heavy investment which both groups had undertaken in the U.S., 'a full and careful consideration of the future oil policies of the British Dominions and Crown Colonies becomes . . . a matter of urgency'.[8] The need to appease American desires and to open up the Crown colonies and Dominions to American enterprise led to the belief among certain sections of the British government that the 'exclusion of foreigners from the control of oil enterprise in British territory [was] unsound in principle, and [had] very little practical advantage'.[9] H.G. Chilton, Sir Auckland's successor at Washington, did not share this opinion, and felt that the threat to *Shell*'s interests in the U.S. had been exaggerated 'for the reason that small owners of oil lands have a political interest in maintaining the Royal Dutch Shell group in the field as a competitor of the Standard'.[10] By eliminating *Shell, Exxon* would be left as the only major determinant of property prices. This certainly was not the case, and exaggerated *Shell's* position in the U.S. compared to *Exxon*'s. Moreover, other major U.S. oil companies would still ensure sufficient competition for *Exxon*. Chilton was right, however, in asserting that the position had been overstated. Although President Harding's first Interior Secretary, Albert B. Fall, denied Creek Indian oil leases to a *Shell* subsidiary, this decision was reversed by Mr. Wock, his successor.

Moreover, the oversupply of domestic oil stilled the public debate of American participation in foreign countries for a while. Chilton wrote:

> Public feeling in this country with regard to our policies may be said, therefore, to be quiescent for the time being. It was never at any time strong enough to justify our pandering to it by throwing away any real advantages, actual or prospective, which we might derive from keeping the oil resources of the Empire under British control, and which outweigh the difficulties and anxieties of the Royal Dutch Shell group.[11]

Despite this, the U.S. government's 'open door' policy showed 'no sign of cessation',[12] and it was necessary for the British to re-formulate their existing oil policy to take into account American feelings.

This, however, would take some time. Clarke of the Petroleum Department felt that British oil policy had been unsuccessful in developing the Empire's oil resources. The Colonial Office countered that the Empire's small production was a good defence measure because resources would be conserved and not exhausted. But Clarke pointed out that undiscovered oil sources were of little value in an emergency. The Colonial Office further contended that foreign interests could, through a minority interest in British concerns, exploit the Empire's oil resources. Clarke, however, felt that no 'American oil companies will wish to participate on these terms',[13] and therefore concluded that there was no harm in allowing foreigners in on an equal basis with British companies. He reasoned that 'if foreigners like to come and spend money in looking for oil, it seems foolish to prevent their doing so: if they were successful, the oil position of the Empire would be strengthened'.[14] The arguments in favour of abolishing British control were that foreign capital would assist British firms in developing the oil resources of Empire, that friction and suspicion with regard to British oil policy in general would be minimized, and that retaliation against British policy could injure British interests in the

U.S. and elsewhere.[15] Within the Board of Trade and the government in general, there was little acceptance of these recommendations because it was confidently expected that the Empire, given
time, would yield greater oil production than hitherto had been the
case. The Admiralty expressed the same fears which prevailed in
1904, and argued that the need to protect the Empire and
consequently the need to ensure oil supplies were an ever present
factor in formulating policy. If the Empire were opened up to
foreign companies, mainly American, they would monopolize the
concessions and prevent their development. However, with the
incorporation of American interests in *TPC,* and the American
Petroleum Institute Committee's finding in 1926 that U.S. oil
resources were not being exhausted, the immediate demands for an
'open door' policy diminished.

Nevertheless, it still remained a source of Anglo-American
friction which would be used by the U.S. to influence British policy
objectives to the detriment of other British interests. Moreover, the
U.S. still remained the largest UK oil supplier. The Middle East,
the cornerstone of the plans for British oil independence, failed to
increase its production substantially, so Britain's dependence on
U.S. oil was not lessened. The Empire's own production had
steadily decreased since the end of the war from 2.5 per cent of
world production to 1.6 per cent in 1927, distributed as shown in
Table VI.

TABLE IV

Distribution of the Empire's oil production, 1926–27 (in million tons)

Colony/Dominion	1926	1927
India	1.2	1.1
Trinidad	0.7	0.8
Sarawak	0.7	0.7
Canada	0.05	0.06
TOTAL	2.65	2.66

Source: POWE 33/253 Board of Trade (Petroleum Department), 'Petroleum
Industry in the British Empire', Feb. 1928.

According to the Petroleum Department, only one-tenth of this
production reached the British home market. In 1926 total Empire
oil imports into the UK amounted to 72.5 million gallons, while
foreign oil accounted for 1.8 billion gallons.[16] The depressing aspect
was that there appeared to be no 'possibility of a considerable
increase in the Empire's production in the future'.[17] Production was

constant in the Burma oilfields, the largest in the Empire, until 1928 when it began to decline. Apart from Trinidad and Sarawak, where production was around 700,000 tons annually,[18] 'the indications for finding petroleum are by no means favourable'[19] in other parts of the Empire. The Board of Trade on 20 August 1926, appointed a committee to 'advise upon questions connected with the economic use of fuels and their conversion into various forms of energy, having regard to national and industrial requirements, and in the light of technical developments'.[20] As a result the government-funded Fuel Research Station at Greenwich started to investigate the hydrogenation of coal and tar into oil. The results were so encouraging that Dr. Lander, the Director of the Fuel Research Station, stated at the Imperial Conference held in London in 1930 that 'there are reasonable grounds for supposing that petrol of very good quality could be produced by this means at a price that would not be too expensive'.[21] The Conference adopted the resolution presented by the General Economic Committee on Oil, which was:

> The Conference notes with satisfaction the progress which is being made in connection with the question of the extraction of oil from coal and the development of geophysical methods of surveying, and recommends the Governments concerned to support such steps as may be found practicable to promote the searching for and production of natural oil, and to increase the supply of refined oil produced within the Commonwealth, whether by the refining of oil or by the production of oil from coal.[22]

The British government had tried to encourage domestic production of shale oil when in the Finance Act of 1928 it levied a four pence custom duty on imported motor spirit but exempted local production. Encouraged by this the Scottish shale industry adopted measures to 'increase production of motor spirit by cracking some of the heavier products, thereby obtaining a larger yield of spirit'.[23] But although there were further increases in the duty levied on motor spirit in 1931 to eight pence per gallon, shale oil production declined from 2 million tons in 1928 to 1.4 million tons in 1936.[24]

Venezuela, where *Shell* was the largest oil producer up to 1934,[25] appeared in the mid-twenties likely to become one of the world's largest producers. As we can see from Table V given below, Venezuela in 1928 displaced the Soviet Union as the second largest oil producer in the world after the U.S.

TABLE V

Percentage distribution of world crude oil production, 1920-30.

Country	1920	1928	1929	1930
USA	64.0	67.2	66.9	62.4
USSR	3.7	6.4	6.6	9.6
Venezuela	0.07	8.4	9.9	10.2
Romania	1.0	2.4	2.3	2.9
Persia	1.0	3.2	2.8	3.1
Mexico	24.0	3.9	3.3	3.0

Source: Adapted, 'Producción mundial petrolera en 1920', *Boletín del Ministerio de Fomento,* 2:16 (Enero 1922), p.12; and, POWE 33/461 Petroleum Dept., 'International Control of Petroleum', Oct. 1932.

The oil companies were attracted to Venezuela because of the relative political stability which the Gómez dictatorship offered, and the good terms for the exploitation of the country's oil resources. Unlike the Middle East or Iran, Venezuela devised a concessionary system whereby most oil companies regardless of nationality could operate, and production costs were much lower than in the U.S. The number of barrels produced per active well was much higher than in any American oilfield (seen in Table VI); moreover, 80.5 per cent of the total number of wells drilled up to 1935 were productive, demonstrating much lower exploration and drilling costs in Venezuela than in the U.S.[26]

TABLE VI

Number of barrels produced per active well in Venezuela and the U.S. in 1928

Region	Production (million barrels)	No. of active wells	barrels/active wells (000s)
Texas	255.4	910	280.6
Oklahoma	247.5	715	346.2
California	232.0	511	455.0
VENEZUELA	106.5	85	1,253.0
Kansas	38.2	225	169.6
Arkansas	32.1	62	518.4

Source: Adapted, E.B. Hopkins, 'Some Geologic and Economic Notes on the Venezuelan Oil Developments', Paper given at the American Association of Petroleum Geologists, Fort Worth Meeting, March 1929, US National Archives, Department of State (DS) 831.6363/149.

Up to 1934 the *Shell* Group of companies dominated oil production, but two companies, *viz Exxon* and *Gulf Oil*, started in 1925 to make inroads into this dominance and by 1928 their joint production totals surpassed *Shell*, controlling 60 per cent of total production by 1936. The three companies dominated the industry accounting up to 1936 for 99 per cent of total production.

Venezuela's crude, a great part refined at *Shell*'s large refinery at Curaçao under Dutch colonial rule, was seen as an important new alternative source of oil for Britain.[27] The changing pattern of UK oil supplies given in Table VII shows that by 1930 American and Mexican oil supplies decreased in importance. Venezuelan crude

TABLE VII

The changing pattern of U.K. oil supplies, 1900–39 (per cent)

Country	1900	1910	1920	1930	1939
USA	61	75	61	34	23
Russia & Romania	37	16	1	17	5
Far East	2	8	5	3	2
Middle East	–	1	5	23	21
Mexico	–	–	28	6	–
Venezuela & others	–	–	–	17	49

Source: J.D. Butler, 'The influence of economic factors on the location of oil refineries (with primary reference to the world outside the USA and USSR)', *The Journal of Industrial Economics,* 1:3 (July 1959), 187-201, Table 2, p.190.

displaced completely Mexican crude,[28] while Middle Eastern oil, together with Russian and Romanian supplies, diminished America's dominant position to a large extent. Venezuela's geographical position placed her at an advantage over both her Mexican and Middle Eastern competitors. Maracaibo, the main Venezuelan oil port, was 113 miles nearer New York than Tampico and 863 miles nearer Southampton than the Mexican port, and only 644 miles from the Panama Canal. This meant that it was 'favourably situated for export to Europe, to the United States and to the Pacific Coast of America, and the Far East via the Panama Canal'.[29] Her logistic production problems were far fewer than her Middle Eastern counterparts. The imposition in 1932 of the U.S. oil import tariff further strengthened Venezuela's position in the British and European oil markets, because the tariff allowed bonded crude oil to be refined and re-exported from the U.S. The effect was for Venezuelan crude to replace part of U.S. domestic oil exported to Europe (see Table VIII). The tariff also made certain companies operating in the country, notably *Standard Oil Co. (Indiana)*, which

TABLE VIII

Percentage change of crude oil imports into Europe, 1928–33.

Year	USA	Venezuela	Romania	USSR	Persia
1928	38.8	13.4	7.8	8.4	14.2
1929	34.8	12.5	8.5	10.3	13.7
1930	33.5	13.0	10.3	12.3	13.6
1931	27.3	14.5	12.5	14.4	13.4
1932	21.5	15.3	13.4	15.8	13.6
1933	18.6	21.4	13.6	11.1	13.0

Source: Adapted, 'Shifts in European supply and the Iraq Oil', *Petroleum Press Service*, 1:15 (Aug.1, 1934), 1-3, p.2.

had previously sent their oil to the U.S., dispose of their considerable assets in the country to *Exxon*, because they did not possess adequate marketing facilities in Europe. Finally, the move in the early 1930s to locate refineries near market areas rather than near crude oil sources increased demand for oil in Britain. By 1932, therefore, there had been a radical change in Britain's oil supplies, with Venezuela becoming the single most important source as we can see from the Table IX given below supplying on the eve of World War II 40 per cent of British oil needs.[30]

At the end of the war *Shell* was favourably situated to reap the rewards of the prescient decision taken by Deterding in 1913 when in the 'most speculative venture'[31] of his life he acquired a number of concessions in the country. By 1928 Venezuela overtook both

TABLE IX

1932 Percentage of total U.K. imports from each of the principal oil-producing countries as compared with ten years ago

Country	1932	1922
Venezuela	23.0	3.9
Persia	21.8	24.9
USA	20.8	33.9
Romania	10.0	1.2
USSR	7.4	0.7
Mexico	7.3	30.2
Dutch East Indies & Sarawak	2.7	1.2
Trinidad	2.1	2.5
TOTAL	94.1	98.5

Source: CAB 50/5/Secret/O.B.122, Committee of Imperial Defence, Oil Board, 'Eighth Annual Report', 31.7.33.

Mexico and the U.S. as *Shell*'s largest single producer of oil.[32]

Venezuela's oilfields were also conveniently situated near sea transport, with easy access to the large American and European markets, and the offshore Dutch islands of Curaçao and Aruba provided a safe haven to build large refineries.[33] The development of the Venezuelan oilfields offered the companies the flexibility they longed for, *viz*, to be able to adjust output and distribution to changing world markets. As a result, after the war, *Shell* was joined by the large American companies, such as *Exxon* and *Gulf*, the former displacing *Shell* in 1934 as the largest single producer.

The predominance of Venezuela as a major oil producer was closely linked to the American domestic oil industry, as well as the need to supply the growing British and European oil needs. The desire of the large integrated oil companies to control the American oil industry stimulated the development of Venezuelan production, and these factors had a direct bearing on British oil supplies.

THE U.S. OIL INDUSTRY

The unpredictable nature of crude oil supplies in the U.S. during the 1920s and early 1930s, due to the *modus operandi* of the industry, compelled large vertically integrated oil companies to seek means of controlling and stabilising domestic production in order to maximise profits. Foreign oil would be used to supplement and in so doing stabilize domestic production, but more importantly it would be used to supply foreign markets which had hitherto been supplied with American oil. It was therefore apparent that foreign oil would form an increasingly important element in the operation of a large American vertically integrated oil company. Similar considerations also applied to *Shell* because, as we have seen, it had a vested interest in the American oil market. As a result of the U.S. being by far the largest oil producer in the world, what happened in its domestic oil industry was of the utmost significance in determining whether foreign oil sources would be developed. To understand the link between U.S. domestic oil production and foreign oil the American oil industry must be examined in detail.

By law the landowner in the U.S. is also the owner of subsoil mineral rights. In the case of oil the 'rule of capture' prevailed, meaning that 'regardless of where the oil may have been *in situ* it belonged to the owner of the land where it was actually brought to the surface'.[34] This provoked a mad scramble for oil, and production in the short term had no connection with demand. In the long run, however, an increase in price tended to lead to more exploration and greater supply.[35] However, there was no connection between the cost of crude oil production and the price at which it was sold. As Shuman states, this was due to the large number of potential suppliers while the 'number of potential purchasers of crude petroleum especially the leaders in price-posting, is relatively small'.[36] The very nature of the oil industry, then, prevented companies from operating under competitive conditions. The uncertainty of exploration, the 'peculiar relationship of fixed and variable cost',[37] together with the inelasticity of demand for the product in the short run, created very unstable market conditions. The high fixed costs of drilling and equipping a well[38] demanded that as much oil as possible had to be produced in order to recoup outlay. In effect, once new reserves were discovered, production carried on at decreasing costs. U.S. producers thus had a very short time frame in which to operate and did not think of the future effect their action would have on the rest of the industry, *viz* lower prices, thereby creating greater collective losses to all concerned. In Frankel's view the industry was not 'self-adjusting' in the sense that a fall in prices significantly restrained supply or stimulated demand.

Crude oil buying and selling in the U.S. was shaped by the development and administration of refineries and pipelines. The pipelines served to control the volume of oil reaching the refineries, and the refiners' throughput capacity determined the volume of oil supplied to them and consequently the amount of oil carried by the pipelines. The refiners would 'post' the price at which they were prepared to buy oil, and effectively determined the price at which the oil was purchased. This system, however, with its flush and famine cycles, posed a number of problems to the large vertically integrated oil company with its own oil supplies since the demand and supply functions for the crude oil industry and for refined oil products are different. The demand for crude oil is a derived demand for refined products, and because of this the maximum amount of oil which is bought in the short run is determined by the physical capacity of the refining industry, and not by direct consumer demand. In addition, because of the operation of the 'rule of capture' in crude oil production, supply bears no relation to demand, producers having to extract as much oil as possible. In contrast, the refining industry is not under the same pressure as the crude oil one to supply refined products at whatever price.[39] In the long run, crude oil prices tend to follow the general pattern of refined products, but because of the differentiation between markets, low crude oil prices in the U.S. during the 1920s and 1930s stimulated the number of independent refiners entering the market. The high cost of entry due to the initial capital costs of a refinery was offset by the availability of cheap oil. The fall in crude oil prices widened the gap between product and crude oil prices, hence the refiner's profits increased. These lower crude prices meant that new refiners could undersell the traditional suppliers, with the result that the large vertically integrated oil company faced the 'unpleasant problem of minimising losses rather than the pleasant one of maximising profits'.[40] High crude prices, as Shaffer concludes, 'severely limit the profit potential of an independent refiner',[41] especially if he has no crude production of his own, and therefore acts as a barrier to entry for potential refiners. High crude prices, however, do not place any obstacles to the vertically integrated oil company because they 'merely enable him to transfer profits from the refining to the production sector'.[42] The large integrated oil company with its own pipeline facilities had available important profit-sharing and maximising opportunities which would allow it to survive and even prosper in the chaos of the market. Nevertheless, low crude prices in the long run had the effect of lowering the profitability of the entire industry compared to other industries, as we can see from Table X given below.

TABLE X

Net Income of 28 oil producing and refining companies compared with 735 industrial companies in the USA, 1926–36 ($ million)

Year	28 oil producing and refining companies		735 industrial companies	
	Amount	Index numbers 1929=100	Amount	Index numbers 1929=100
1926	571.2	94.7	2,456.3	75.6
1927	255.1	42.3	2,171.9	66.9
1928	537.7	89.2	2,779.9	85.6
1929	602.9	100.0	3,248.9	100.0
1930	265.1	44.0	1,807.2	55.6
1931	72.7	–12.1	681.1	21.0
1932	72.8	12.1	85.6	2.6
1933	101.4	16.8	710.7	21.9
1934	179.7	29.8	1,083.2	33.3
1935	282.0	46.8	1,582.1	48.7
1936	451.2	74.8	2,359.2	72.6

Source: Adapted, US Congress, *Petroleum Industry,* Investigation of Concentration of Economic Power, Part 17, Section 4, Hearings before the Temporary National Economic Committee, 76 Congress (Washington: 1940), Table E, p.7706.

In addition the profitability of the leading 28 oil companies between 1927 and 1935 was lower (except for 1932) than the 735 industrial companies. According to Del Sesto, Special Assistant to the Attorney General in 1940, the recovery in profitability after 1935 was due in large measure to the enforcement of pro-rationing laws which had the effect of pegging crude oil prices at high levels.[43]

The stabilisation of oil prices through production control was seen by the large vertically integrated oil companies as being vital to the survival of the industry. Production control promised higher and more stable refinery margins through its effect on entry by potential competitors. As long as production was uncontrolled, entry into refining, and thus protection of the market for refined products, could not be accomplished. In order to achieve higher prices it was necessary to bring production under control because the destabilising effects of price fluctuations were affecting the profitability of the industry. Both buyers and sellers had much to gain from stability. In time of flush production, for example, sellers could not dispose of their products because the 'rule of capture' produced an 'oil rush' resulting in overproduction, low prices and low physical recovery of oil. The 'rule of capture' also imposed heavier capital costs by forcing each individual producer to extract as much oil as possible. There was thus a common interest in seeking release from this. However, while crude oil producers had a vested interest in production controls, the inherent contradictions of the system ensured that these would not be achieved voluntarily. Most independent producers were wildcatters, so that any curtailment of production would hit them most severely. For the buyers of crude oil, price stability meant better development of the industry because they were able to plan for the long-run needs of the industry.

The oil industry is a capital intensive industry, as we can see from the Table XI given below.

TABLE XI
Capital intensity in the United States petroleum industry

Division	Annual wages ($ millions)	Investments ($ millions)	Ratio of Investments to wages
Exploration & production	232	6,493	28
Pipeline transportation	50	1,104	22
Refining	165	3,718	22.5
Marketing	559	3,210	5.7

Source Raymond F. Mikesell & Hollis B. Chenery, *Arabian Oil* (Chapel Hill, 1949), Table 7, p.155.

The typical ratios of investment to wages for the American manufacturing industry at the time were 5:1. The effect of high capital intensity in the various sectors of the oil industry is that any expansion in output involves a large initial capital outlay, causing the large oil companies to look further into the long term when planning their operations. Consequently, a smooth growth pattern could only be achieved if the domestic crude oil market was stabilised. This was achieved by fostering the belief in the necessity to conserve oil supplies in the U.S. As a result a number of pro-rationing laws were put into effect which restricted supply to demand, thus ensuring a high price schedule for crude oil.

The acute problems of oversupply together with the 1924 Teapot Dome scandal[44] brought an awareness of the need for greater government control of oil production to ensure smoother supplies and conserve oil. During the 1920s and 1930s there was a gradual adjustment of oil men and government to Federal and State regulations of the industry, drawn up as a result of the industry's economic structure. In December 1924 President Coolidge appointed the Federal Oil Conservation Board to consider and recommend a national policy for the oil industry in view of the need to conserve oil for future needs. It reported in 1926 that the States should co-operate to pro-ration oil production, and that foreign markets held by American oil companies should be supplied with foreign oil.[45] The oil states of Oklahoma, California and Texas co-operated for a while in restricting production by voluntary agreement, but the discovery in 1929 of two large oilfields in California and East Texas brought to an end these efforts. In June 1929 oil men together with Federal and State government officials met at Colorado Springs to discuss plans for an interstate compact.[46] Although the Independent Oil Producers stormed out of the meeting in protest, a Code of Fair Practices, under the Federal Trade Commission, came into being. However, further domestic oil discoveries and increasing foreign imports undermined the effectiveness of the Code. In 1932 a tariff was placed on foreign oil imports, and in the following year a further Code of Fair Competition for the Petroleum Industry was enacted under the provisions created by the National Industrial Recovery Act of the same year. The Code restricted domestic production of oil to demand, as well as reducing imports to an average of the last six months of 1932.

The aggregate effect of the various means to curtail oil supplies, together with the Depression (which had the result of drying up investment), was that domestic production and imports fell during 1930–32. Although prices dropped initially (reaching its lowest level in July 1931 of $0.24 per barrel), by the end of 1933 they had stabilised at the relatively high level of $1.00 per barrel. In addition, oil price shifts were reduced from 100 per cent to 10–20 per cent during one year.[47] All this benefited the large oil companies for as output per well was restricted the timescale to amortise the investment lengthened, thereby increasing the possibility of a small operator with limited capital going bankrupt (since he could not increase his revenue by increasing production), and thus allowing the large companies the opportunity of acquiring the properties cheaply.[48] But major companies were not only interested in restricting supply in order to lessen competition from small oil producers. A large number of producers with high production rates indicated that oil prices were low, which in turn encouraged Independent refiners to compete more vigorously with major refining companies. This competition would be restricted and kept in check by maintaining relatively high crude oil prices. Some Independents, therefore, viewed the code with scepticism, seeing it as a way of perpetuating scarcity and monopoly, and a number of lawsuits were brought alleging that the code was unconstitutional.

The Schecter decision of the Supreme Court on 1 July 1935, ruled that the National Industrial Recovery Act was unconstitutional, and consequently Federal pro-rationing of oil came to an end. But by this time prices had increased and the industry as a whole was less panic-stricken and better equipped to face new challenges. In the same year Congress authorised the Interstate Compact to Conserve Oil and Gas which formalised the working relationship between oil-state governors and oil companies that existed prior to the National Industrial Recovery Act. Under the guise of conservation the Interstate Compact ensured that production never outran demand, that is, 'that the amount of oil the industry leaders estimated is likely to be consumed at indirectly-set prices'.[49] Although the Compact endeavoured to regulate and control the industry, real power still remained in 'the hands of the giant oil-purchasing companies for more substantial reasons'.[50] The major companies controlled much of the production, pipeline, refineries and markets, and viewed oil operations on a global scale rather than a national one. Pro-rationing, then, brought about a more orderly marketing of crude oil with fewer and less extreme price changes. This price stability was accompanied by a rising trend which raised the initial value of investment as the ultimate recovery of oil was higher, although restrictions of output from new wells operating on natural drives and the 'consequent extension of time for ultimate recovery either of cost or value . . . would tend to lower present value'.[51] As Chazeau and Kahn suggest, the most important economic impact pro-rationing had was that it destroyed the 'nexus between oil discoveries and short-run price'.[52] It was not surprising, therefore, that major oil companies already committed, and appreciative of the advantages of integration, decided to improve the balance between their upstream and downstream operations.[53]

The economic advantage for an oil company of vertical integration is that it is assured an outlet for its crude production, leading to smooth cost-effective planning schedules. Its refinery can be programmed to take advantage of the most efficient use of equipment. The company is also capable of a more flexible and efficient adjustment to short-run changes in demand for different products in different areas, which can be quickly reflected in the inflows of crude oil. A further incentive is that the company is able to shift costs of production from various areas of the world or of its integrated production processes in order to pay the least amount of tax, and consequently achieve the highest possible revenue.[54] A smooth flow of oil ensures that prices remain stable and hence the company is better able to plan for the future. However, since a 'drop in output will raise their costs more proportionately than in almost any other industry, market sharing and other methods of assuring a stable state of output are particularly attractive to the international oil companies'.[55]

The oil industry outside the U.S. is differently structured, and is effectively organised into a strong oligopoly. Here Adelman argues that the industry is self-adjusting because expected oil price rises are met by an increase in production and vice versa.[56] As long as marginal cost is less than anticipated price there is an incentive to produce. Production is also not subject to the uncertainty of crude oil exploration which was the source of glut and scarcity in the U.S. In foreign countries the discovery of new reserves does not constitute new supply. It is the producers' expectation of profit-making that will determine whether the field is developed. The costs of developing the discovered reserves, then, are the primary factors preventing unprofitable oil from flooding the markets. Nevertheless, heavy investment is needed to ensure adequate sources of crude 'in advance of requirements to facilitate planning for the maximum use of facilities in their higher stages, and to

protect the investment in the highly specialised equipment at the
upper levels'.[57] A further incentive for producing foreign oil was
that the vertically integrated oil companies could supply their
European and American markets with cheap foreign oil, charging
high U.S. oil prices which the conservation policies of the mid-
thirties produced. This was possible because the international price
structure for oil was set by the U.S. Gulf-area oil prices.

The international price structure of oil was based on the price of
crude being fixed with the 'posted' prices of the Gulf-area of the
U.S. The result was that European and world independent refiners
paid the high 'posted' price of the U.S. Gulf-area for cheap
Venezuelan and foreign oil.[58] This system allowed the vertically
integrated oil companies to act as true oligopolists, and to protect
their investments in the U.S., as any price cutting by foreign oil
countries would have undermined the American domestic industry.
The Gulf + system as it came to be known ensured that the f.o.b price
of oil in any part of the world was determined by Gulf prices,
thereby eliminating differences in prices from various supply
sources. The price paid by importers, however, varied according to
the actual destination of oil because of differing freight charges. The
companies were therefore able to maximise their global operations
by supplying the world market from the source that gave them the
best return, while consumers the world over paid the same price
wherever the oil came from. There were hidden benefits in these
practices because the companies saved on transport costs, and did
not pay the relatively higher U.S. labour rates for domestic coastal
shipping.[59] The higher Gulf oil prices were, the more profitable it
was to develop cheap foreign oil sources. As a corollary to this the
conservation measures introduced during the 1920s and 1930s,
which led to a higher price level for crude oil, also induced the large
vertically integrated oil companies to develop foreign oil sources. In
addition the international price structure of oil, maintained at
artificially high levels with pro-rationing, had the effect of
preventing Venezuelan crude (and later Middle East oil) from being
a disturbing factor in the U.S. domestic scene.[60] Low cost oil
producing countries were effectively barred by this system from
competing in terms of price advantage against each other or with
domestic U.S. oil.[61] The Gulf + system guaranteed against a price
war for under normal competitive conditions, foreign oil suppliers
would have a greater advantage because of lower costs, location, and
other minor factors, than their counterparts in the U.S., and would
inevitably secure a greater share of world oil markets. Moreover, as

Dirlan argues in the case of Middle East production, high crude oil prices tended to keep foreign oil production far below its maximum efficiency recovery rate, and in so doing prevented it from flooding the market. The system prevented governments from stimulating production and assured that production in the underdeveloped countries was determined by the requirements of the vertically integrated oil companies in the large European and American markets. In order to safeguard against any one company or country upsetting the system the major oil companies entered into a series of agreements to limit competition and retain their respective market shares.

THE INTERNATIONAL PETROLEUM CARTEL

The only way that the world oil industry could be stabilised would be for the large vertically integrated oil companies to combine in a marketing and producing cartel to control world oil supplies. Deterding with little success had tried to reach such an agreement with *Exxon* in 1910 to limit price wars. The latter could offset any losses incurred in the price wars with profits from its American operations. In 1911 Deterding therefore started 'laying plans to commence both producing and marketing operations in the United States, seeking to form an anti-Standard bloc there under cover of the confusion created by the dissolution decree'.[62] He also tried to build up a strong European concern to combat *Exxon* on equal terms, hence his desire to acquire control of *Anglo-Persian* and its vast potential oil supplies from Persia. As we have seen this was almost accomplished soon after the end of World War I but the terms were unacceptable. However, the experience in forming *TPC* served to demonstrate that *Shell* and *Exxon* could work together. But by 1926, a price war had broken out between *Shell* and *Standard Oil Co. (New York)* in India. Both companies bought Russian crude, refined it, and with it supplied the Indian market and others in the Far East. After several years of unsuccessful negotiations to regain its Russian properties nationalised in 1918, *Shell* decided to retaliate by placing a boycott on Russian crude and asked *Standard Oil Co. (New York)* to join. *Shell* had no trouble in doing this because it had Romanian oil sources to keep its Indian refineries working, whereas *Standard Oil (New York)* had no alternative source to replace Russian crude, and so declined to join in the boycott. Annoyed by this, Deterding announced on 19 September 1927, that the price of kerosene would be reduced in India if

Standard Oil (New York) did not curtail its Russian oil imports. The American company retaliated and prices began to fall, with the price war later spreading to the U.S. and Europe. This episode made it clear that 'control of reserves in newly discovered fields was not, in and of itself, an effective guarantee against the outbreak of competition'.[63] It was necessary to reach a working agreement with the major producing oil companies. Events in the German chemical industry at the time only served to reinforce this view more strongly.

In 1869 M. Berthelot, a French chemist, made the first laboratory tests to obtain oil from coal. Twenty-one years later Parker obtained similar results in England, and during World War I, Herr Noelleddorf, War Director of *Deutsche Erodel A.G.*, supplied German U-boats with synthetic oil produced from coal. The costs of producing such oil were very high. However, after the war, Professor Bergius of Heidelberg University perfected a hydrogenation production technique that produced oil from coal in commercial quantities at competitive prices, using Berthelot's method. At the World Conference of Coal held in Pittsburgh in 1925, Bergius announced his technique, claiming that the 'process was capable of converting coal into gasoline or any other oil products. It was also capable of converting crude oil directly into gasoline 100 per cent by volume, and of producing high-grade oil products from low-grade crude'.[64] Before the war Bergius had patented his technique world-wide. After the war, however, Bergius' U.S. patents were held by the Chemical Foundation, the transferee of the Alien Property Custodian. The patents outside the U.S. were held by the *International Bergin Co.*, the controlling interests of which were held by *I.G. Farben, Shell*, and *Imperial Chemical Industries (ICI)*. Later, *I.G. Farben* patented several further processes related to the hydrogenation technique.

When it made enquiries in Germany about the hydrogenation process *Exxon* was shaken to discover the potential threat that the new process posed to its business. On 28 March 1926, after conferring with *I.G. Farben* officials, Frank Howard, an *Exxon* executive, wrote to Teagle, President of *Exxon*, from Mannheim stating that:

> Based upon my observations and discussions today, I think that this matter is the most important which has ever faced the company since the dissolution. The Badische can make high-grade motor fuel from lignite and other low-quality coals in

amounts up to half the weight of the coal. This means absolutely the independence of Europe on the matter of gasoline supply. Straight price competition is all that is left. They can make up to 100 per cent by weight from any liquid hydrocarbon, tar fuel oil, or crude oil. This means that refining of oil will have as a competitive industry in America and elsewhere catalytic conversion of the crude into motor fuel.[65]

This led to a series of meetings between the German chemical company and *Exxon*[66] to reach an agreement. Both companies sold oil in Germany, but *I.G. Farben* did not have the marketing facilities to compete domestically or internationally with the major oil companies. Although the hydrogenated oil process competed successfully with prevailing prices, it was likely that the oil companies, especially *Exxon*, would use severe price undercutting in any market which *I.G. Farben* entered to drive them out of business. The battle to secure an increase in their market share would therefore be extremely tough and a tremendous drain on *I.G. Farben*'s resources. Moreover, the oil companies could easily enter the chemical industry, posing a threat to *I.G. Farben*. Securing an agreement with *Exxon* over oil in Germany to ensure that it kept out of the chemical business was, therefore, in their own interest. *Exxon*, for its part, also wanted to reach an agreement to avoid further price wars, and because its arch-enemy *Shell* held an interest in the *International Bergin Co.* which in turn retained all the world patents for the hydrogenation process, it was not inconceivable that *I.G. Farben* might join *Shell* (which had all the marketing facilities at its disposal) to unleash a ferocious price war in Europe against *Exxon*. Consequently, *Exxon* agreed to enter into an agreement with *I.G. Farben* by which it would refrain from selling its own oil but would purchase *I.G. Farben*'s production to market in Germany. On 27 September 1927, a further agreement was reached whereby *Exxon* acquired the right to licence the production of hydrogenated oil in the U.S. under *I.G. Farben*'s American patents.[67]

At first the arrangement did not work because *I.G. Farben*'s position in its home oil market was undermined by the importation of *Exxon*'s own oil into Germany. Several unsuccessful meetings took place between the two companies. On 31 August 1928, Dr. Bosch, head of *I.G. Farben*, declared that

the only solution would be to consider that the two companies were married for a long period and that each must respect the other's interests; that a real solution could be found only by realising that the I.G. interest and the S.O. interest were the same and that whatever Standard did would react to the interest of the I.G. as well.[68]

Moreover, *I.G. Farben* felt that if the agreement was to have any lasting effect *Shell* would have to be incorporated into it. As we have seen *Shell* held an interest in the *International Bergin Co.*, hence *Shell* could exercise its rights and use the hydrogenation patents for its own use. The solution was for the German chemical company to sell the company to *Exxon,* while retaining the German patents. *Exxon* in turn would transfer half the shares to *Shell,* and the patents would then be held by a new company owned by *Shell* and *Exxon.* This meeting led directly to the Achnacarry Castle Agreement which divided up the marketing of the world's oil resources among the major oil companies.[69]

On 17 September 1928, Walter Teagle, Sir Henri Deterding and Sir John Cadman met at the latter's home and signed the Pool Association Agreement, usually known as the 'As Is' or 'Achnacarry Agreement' of 1928. The agreement called for the three companies to pool their resources in order to streamline their world-wide operations. At the heart of the agreement were seven principles which were: (1) the three companies would accept their 'present' volume of business and their proportion of any future increase in consumption; (2) companies would have the use of their partners' facilities; (3) construction of only such additional facilities as were necessary to satisfy an increase in demand; (4) production to retain the advantage of geographical location on the basis that 'values of products of uniform specification are the same at all points of origin';(5) supplies would be drawn from the nearest producing area; (6) excess production over consumption to be shut in by producers in each producing area; and (7) elimination of any competitive measures or expenditure which would materially increase costs and prices. To comply with U.S. anti-trust laws, the U.S. domestic oil market and export trade were outside the scope of the agreement. However, by limiting production of oil in foreign countries, imports of oil to the U.S. would necessarily be limited. The export trade from the U.S. was to be controlled. World pricing for the cartel would use U.S Gulf prices as a single governing basing point. Adjustment for quality and standard tanker charges for U.S. Gulf ports of destination were to be added to the basing point price.

The companies would thus be able to sell cheap foreign oil at expensive American prices. While the Gulf + system remained, it was necessary for the intra-cartel companies to set up their own multiple-basing point system. Through reciprocal exchange of supplies, the cartel members would receive their oil from the most favourably situated oil production centre. The savings in transportation costs would result in additional profits for all companies.

The oil companies also sought to implement the 'As Is' agreement by forming two export trade associations in the U.S. under the Webb–Pomerene Export Trade Act. This would restrict American exports and at the same time bring the domestic oil industry under their corporate control. The *Standard Oil Export Corp.*, controlled by *Exxon* and its subsidiaries, was organised on 26 November 1928[70] for the purpose of channelling all exports through one corporation. It served two functions in *Exxon*'s export trade:

> First, it established a single centralised corporate control over the export trading facilities of the parent company's principal producing and exporting subsidiaries. Second, as the corporate agency exercising with the requirement of the international agreements entered into the parent company.[71]

A year later this corporation acquired the *Anglo-American Oil Co. Ltd.*, *Exxon*'s British subsidiary, which marketed and refined its products in the British market. This was 'directly related, on the one hand, to the operations of the *Export Petroleum Association* in pricing and allocating American exports of petroleum in Great Britain and other areas in which the *Anglo-American* operated'.[72] In early 1929, the *Export Petroleum Association Inc.* was formed by the *Standard Export Corp.* and sixteen other U.S. oil companies in the export business.[73] The Association had the machinery to set export quotas in line with the Pool Division Agreement of 1928 for members which were the most important exporting oil companies. The reason behind its organisation was to control American oil exports and fix f.o.b. Gulf prices, which the principal international companies had agreed should be the base price used in determining the value of oil products of uniform specification in all world markets. However, the company failed because the threat of foreign interference meant that members were unable to agree on prices. It was felt that *Shell* could wreck the plans laid out through the *Shell Union Oil Corp.*

With the collapse of the *Export Petroleum Association* there were rumours of price wars, but these were averted by the close collaboration of the member companies. Already, *Shell* and *Exxon* in the East Indies and Asia were working closely together. It was evident though that world-wide control of production and marketing under the present system was only possible on a 'piecemeal' basis. The first step towards a more efficient control of production and marketing by the three original companies of the Pool Division Agreement was taken on 20 January 1930, when a 'Memorandum for European Markets' was drawn up. In it the three companies agreed to act as a unit to gain world-wide control of the industry. They resolved that this could best be achieved by reaching agreement on the control of production in particular producing areas, and by setting up marketing agreements in particular consuming countries. Other companies would be allowed to join, and the agreement would cover all oil products with quotas allotted for each of the product classes as part of the internal trade of the country. The base period for calculating proportions of each market to be allocated to a company would be determined by its particular share in 1928. The members could only expand their quotas at the expense of outsiders. Between 1930 and 1932 it proved difficult to admit outsiders and to remain faithful to the 1928 agreement because of the numerous problems created in sharing markets, fixing prices and making the necessary adjustments. A.C. Veatch, a Director of *Exxon*, wrote in July 1931 that the companies were having difficulty in finding outlets, thus 'instead of competition for the control of oil supplies involving aims of actions productive of international friction, we are dealing at present with the more simple matter of competition for the disposal of a commodity of which for the time being there is an oversupply'.[74] On 15 December 1932, at a meeting held in Texas attended by *Shell, Exxon, Gulf, Socony-Vacuum, Anglo-Persian, Atlantic Refining Co.,* and the *Texas Corp.,* the Heads of Agreement for Distribution agreement was signed, giving greater elasticity to the Pool Division 'by setting out in detail the manner in which the participants to the main agreement would contribute from their quotas in local markets to make room for larger production quotas granted to outsiders'.[75] During the next eighteen months little happened among the oil companies because uncontrolled f.o.b. Gulf prices were very low.

Independent producers in the U.S. increasingly viewed foreign, mainly Venezuelan crude, with great anxiety as they felt that this oil would allow the large vertically integrated companies greater control and manipulation of the market, and consequently pressed for a tariff of $0.42-1.00 on foreign oil. As a result of this pressure the U.S. Tariff Commission began to investigate whether foreign oil enjoyed a competitive advantage over domestic oil. In 1932 it reported that the average cost of American domestic production delivered to the Atlantic Seaboard for 1927–30 was $1.90 per barrel (of which $1.09 was cost of production, $0.04 purchasing commission, $0.49 pipeline costs, and $0.265 tanker charges), while the average cost of Maracaibo crude at Eastern Seaboard ports for the same period was $0.87 per barrel (of which $0.62 was cost of production, $0.25 tanker charges), and the average cost for the rest of the world including Venezuelan oil delivered to the Eastern Seaboard for the same period was $1.15 per barrel ($0.87 cost of production and $0.28 tanker charges).[76] The difference between costs of foreign and domestic produced oil appeared to be enormous, $1.03 and $0.75 per barrel for Venezuelan and foreign oil respectively. However, in order to make a fair comparison, adjustments have to be made; for example, pipeline charges accounted for a quarter of the cost of domestic crude ($0.49), and this was the price charged and not the actual cost of transporting the oil. On the other hand, Maracaibo oil was transported for a short way by pipeline, and the charge in the cost schedules was the actual cost and not the price charged. Consequently, in this respect, U.S. domestic oil was slightly overvalued. But the greatest adjustment had to be made in quality comparisons. The gross value of products derived from one barrel of U.S. crude and one barrel of Maracaibo oil was $2.74 and $1.71 respectively, but refineries using domestic crude incurred greater expense in obtaining the more valuable products than the users of foreign oil, with the result that the cost of refining one barrel of domestic oil was $0.73 compared to $0.30 for foreign oil. By deducting these costs, the net realisation from domestic crude was $2.01 against $1.41 for foreign oil, so that foreign crude was worth only seven tenths of U.S. domestic oil.[77] Taking into account the lower yield in gas and fuel oils from foreign oil, the 1927–30 averages for the true cost of one barrel of U.S. domestic oil was $1.33 (70 per cent of $1.90) compared to one barrel of Venezuelan oil at $0.87. This still gave Venezuelan crude a cost advantage of $0.46, though for foreign crude as a whole the advantage was reduced to $0.19.[78]

This cost advantage was reflected in the profit margins of the companies using the different types of oil. Taking 1930, for example, refineries using domestic oil made a profit of $0.11 or 6 per cent per barrel of oil, whereas refineries using Venezuelan oil made a profit of $0.54 or 38 per cent per barrel.[79] These benefits accrued to the large vertically integrated oil companies, but the average delivered price of Venezuelan oil at the Atlantic Seaboard for 1927–30 for the Independent refiners was competitively priced at $1.11, while foreign oil as a whole was priced at $1.35 per barrel. This competed very well with the average 'posted' price for 1927–30 of $1.22 at the well-head. The other advantage of Venezuelan oil for these refiners was that the price included transportation to the Atlantic Seaboard, and for this reason production costs at the Atlantic Seaboard were lower for foreign oil than for domestically produced oil. But the large vertically integrated oil companies did not want foreign oil to undermine the American oil industry but rather, as S.A. Swensrud, President of *Gulf Oil* declared in August 1949, to supplement 'our domestic petroleum supply'[80] in times of shortage, thus preventing the wild fluctuations which the industry had been subject to during the 1920s. Nevertheless, the Senate in June 1932 approved a tariff on imported oil consisting of $0.21 per barrel of crude oil and miscellaneous derivatives, $1.05 per barrel of petrol and other motor fuel, $1.68 per barrel of lubricating oil, and $0.01 per pound of paraffin and other petroleum wax products.

Despite the tariff, during 1932 foreign oil still entered the U.S. in large quantities because the Treasury 'decided that crude oil in bond, to be used in making refined products for exports and fuel oil to be used as supplies for ships engaged in foreign trade could be imported free of tax'.[81] Imports of petrol, however, were negligible after the imposition of the tax.

As we saw previously with the National Industrial Recovery Act Code of Fair Competition for the petroleum industry, Gulf prices advanced from $0.50 per barrel in July 1933 to $1.00 per barrel in October of the same year, and stabilised at that level. The big three companies continued to increase their hold on Venezuelan, Romanian and Far East markets. Past experience had demonstrated that it was difficult to comply with the principles of the Pool Division because the 'diversity of interest of independents with whom these companies necessarily had to deal in local markets was such as to necessitate regarding those principles as ideals to be attained as far as practicable, but not to be insisted upon too strongly'.[82] In June 1934, the seven oil companies signed a new agreement, known as the Draft Memorandum of Principles, which was essentially a restatement of the Pool Division Agreement, in which the following points were agreed: (1) local agreements to which outsiders were admitted were to be regarded as separate; (2) in any market, where all parties to the main agreement operated or might subsequently become members of a local cartel including outsiders, the local cartel would override the Memorandum for that market for the duration of the local agreement; (3) two types of quota for the participants were to be set up in each local market, one covering the total quantity supplied, with each participant including sales to other participants and approved outsiders, and the other covering only quantities actually distributed by each participant; (4) fines were to be imposed against excessive overtraders, consisting in apportioning back sales to the undertraders on a more equitable basis than provided for under the previous agreements; (5) quotas were to be adjusted among traders as a penalty against undertraders, when new members were admitted or when a participant retired voluntarily from the market; (6) only under certain conditions would prices be fixed by majority vote of all the participants in each local market, but even at such times, the practice of open price reporting and discussion of proposed prices in advance was to be followed; and (7) no competitive expenditure for advertising and sales promotion and for capital investment for facilities was to be made except in accordance with budgets previously approved by the cartel's London Committee.[83]

The case of Venezuela, the largest oil producer outside the U.S., illustrates well how the agreement operated among the three largest oil companies, *Shell, Exxon* and *Gulf Oil,* to maintain production in line with their own needs. The use of 'crude-oil supply contracts to control the production of oil had effectively bound together the major oil interests in Venezuela'.[84] *Standard Oil Co. (Venezuela)*[85] and *Mene Grande Oil Co.,*[86] one of *Gulf Oil*'s subsidiaries in Venezuela, had entered into production agreements on 30 November 1933, and 23 September 1936, and according to the Federal Trade Commission:

> such contracts were of a far more comprehensive nature, welding the interests of the parties together in an explicit joint enterprise lasting for the life of the concessions. Under this agreement the price paid for the purchasers, except for the sum initially paid as a consideration for the agreement, was merely the actual costs of production of the oil. The contracts, therefore, were again devices for sharing the ownership of the oil. Another important feature of the contracts was certain controls that were laid on production.[87]

In 1937 99 per cent of Venezuelan production was directly controlled by the three companies mentioned above. The discovery

of the Eastern Venezuelan oilfields owned by *Exxon* and *Gulf Oil* in
1936 posed a difficult problem for the companies, because they did
not know where to place the oil in the international oil market. Their
own Venezuelan production was 17 per cent below capacity and due
to the Pool Division Agreement, world markets were already
divided. The U.S. did not offer an outlet because of tariffs and the
need to keep the market free from a flood of imports. It was
therefore necessary for the three companies to reach an agreement
over the new source of production. In 1937 and 1938 the companies
entered into a series of agreements to eliminate *Gulf Oil* as an
independent factor in Venezuela. *MGO* would be transformed into
a joint venture enterprise owned and controlled by *Gulf Oil, Shell*
and *Exxon*. The future development and output of Venezuela
would be regulated 'so that a "fair" relationship would be maintained at all times with total world production, outside of the
United States, and to divide the *allowable production* among the
participants to the agreements'.[88] This was to be done by merging
SOV's eastern division with *MGO*, and on 15 December 1937, both
companies signed the first agreement to regulate and bring up to
date the earlier ones. Its main purpose was 'declared to be the
achievement of "economy operation" of certain concessions owned
by the two parties in Eastern Venezuela'.[89] The companies pooled
their property together,[90] and:

> It was agreed that the party owning title to and operating each
> pooled concession or concession interest would transfer one-
> half of its production or share of production to the other.
> Thus, regardless of which party held title to concessions
> which subsequently should prove to be oil producers, both of
> the oil companies would share equally in the oil produced on
> the pooled concessions.[91]

While ownership of pooled property remained undisturbed by the
agreement, 'all rights, benefits, burdens, and obligations arising out
of the ownership of the concessions were shared on an equal basis
between the two parties'.[92] In effect, the two companies became
jointly owned but each operated separately. Although there was to
be a joint management structure this went further than a mere
combination of managerial skills. The accounting manual attached
to the agreement 'suggests that the material and manpower
resources of the two companies were to be pooled in the operation of
the concessions. Detailed instructions are given for allocating costs
incurred by pooled labour, equipment, and professional and

technical services in the developmental projects, the drilling of wells and so on'.[93] On the same day, a separate agreement, known as the *MGO-International (Principal) Agreement,* was signed in Toronto between the *International Petroleum Co.,* an *Exxon* subsidiary, and *MGO* under which *Exxon* agreed to purchase half of *MGO*'s production at cost. *International* also acquired at half price an undivided half interest in *MGO*'s physical capital, and a half interest in all future property acquired by *MGO.* In this agreement *MGO* 'traded away' this equal voice in the 'pooled concessions' and also its freedom of action with respect to its wholly controlled concessions. In so doing, it surrendered to *International* the ultimate and basic direction of operations and the disposition of concessions'.[94] *International* paid for half of *MGO*'s assets by assuming half of *MGO*'s production expenses and by paying $10 million in cash. A quarter of this would be paid by the purchase of oil at $0.07 per barrel, that is just over 357 million barrels over an eight year period starting from 15 December 1937. On the same date three payments totalling $25 million would be credited to *MGO*'s accounts, and similar payments made on the same date in the following two years. Through this agreement the future development of *MGO*'s production would be determined by *Exxon*'s needs. This was also taken into account in an agreement known as The Fourth Party Ratio Agreement, signed by *MGO,* and two Venezuelan *Exxon* subsidiaries, *SOV* and the *Lago Petroleum Corp.,* on the same day as the others in Toronto, and which fixed production quotas between the companies concerned for the next twelve years as well as providing the 'machinery for setting new quotas annually in 1950 and thereafter. These production quotas were prescribed in accordance with certain "principles" and goals which are stated so broadly, however, that the agreement could only have been a partial instrument in their achievement'.[95] On 30 November 1938, *Shell* in an agreement known as the *International-N.V. Nederlandsche Olie Maatchappij-Main Agreement,* acquired one-half interest in the *MGO-International (Principal) Agreement* of 1937, with the result that *Shell* became a full and equal partner with *Exxon* in *MGO*'s production operations. *Shell* paid *International* $50 million cash and agreed to pay half of all current expenditure of *International* in MGO. This arrangement was of great value to *Shell* as the company gained an important foothold in the development of Eastern Venezuela's oilfields.

The merging of the interests of *Exxon* and *Gulf Oil* diminished the rivalry both companies had in supplying Venezuelan oil to

European markets. It also meant that two companies, *Exxon* and *Shell*, controlled the lion's share of Venezuelan oil production, with serious implications for Britain, because by this time Venezuela was the largest single supplier of oil to Britain.

NOTES

1. cf. Ludwell Denny, *We fight for oil* (New York: Alfred A. Knopf, 1928); A. Eugene Havens & Michel Romieux, *Barranca-bermeja. Conflictos sociales en torno de un Centro petrolero* (Bogota: Universidad Nacional, 1966); Ernest Raymond Lilley, *The Oil Industry* (London: Constable & Co. Ltd., 1926); I.A. Manning, 'Petroleum Production – Colombia', U.S. Department of Commerce & Labour, Bureau of Manufacturers, *Monthly Consular & Trade Reports*, No. 332 (May, 1908), 154-55; Arthur H. Redfield, 'Our petroleum diplomacy in Latin America' (Ph.D. Diss., The American University, 1942); U.S. Senate, 'Diplomatic Correspondence with Colombia in connection with the Treaty of 1914 and certain oil concessions', *Senate Doc. No. 64*, 68 Cong., 1 Ses. (Washington, 1942); and, Benjamin H. Williams, *Economic foreign policy of the United States* (New York: McGraw-Hill Book Co. Inc., 1936).
2. cf. Gibb & Knowlton, *The Resurgent Years;* and, Lilley, *The Oil Industry.*
3. cf. Pedro N. López, *Política petrolera* (La Paz: Imprenta Boliviana, 1929); and, Amado Canelas O., *Petróleo: Imperialismo y Nacionalismo* (La Paz: Libreria 'Altiplano', 1963).
4. The first legislation was enacted on December 14, 1907, declaring that the oilfields of Comodoro Rivadavia would be developed by the State, leading later on to the establishment of the *Yacimientos Fiscales Petrolíferos.* Cf. Enrique Mosconi, *El Petréleo Argentino 1922-1930* (Buenos Aires: Talleres Gráficos Ferrari Hnos., 1936), and Arturo Frondizi, *Petróleo y Política* (Buenos Aires: Editorial Raigal, 1956) 2 ed.
5. Peter Seaborn Smith, 'Petroleum in Brazil: A study in economic nationalism' (Ph.D. Diss., The University of New Mexico, 1969).
6. Percentage Distribution of World Oil Production, 1919-1938.

Country	1919	1928	1938
USA	70.7	68.0	61.1
Venezuela	0.1	8.0	9.4
USSR	5.4	6.5	10.5
Mexico	12.7	3.8	1.9
Iran	1.7	3.3	4.0
Indonesia	2.6	2.4	2.9

Source: Adapted, Venezuela, Ministerio de Minas e Hidrocarburos, Dirección General, División de Economía Petrolera, *Petróleo y otros datos estadísticos* (Caracas, 1964), p.142.

7. John K. Towles, 'Petroleum Trade and Industry in the United Kingdom', U.S. Bureau of Foreign & Domestic Commerce, *Trade Promotion Series No. 80,* Supplement to Commerce Reports, 29.1.23.
8. POWE 33/353 Des.371 Geddes (A) to Lord Curzon, 23.3.23.
9. POWE 33/353 J.C. Clarke, 'Minute. Petroleum Policy', 21.4.23.
10. POWE 33/353 Chilton to Warner, 29.6.23.
11. *Ibid.* Exxon entered the Dutch East Indies through a minority interest in the *N.V. Nederlandsche Koloniales Pet. Maatschappij.*

12. POWE 33/353 'Petroleum – Nationality in Oil Leases on Public Lands in British Territory – Statement of the present position by the Petroleum Department', 18.7.23.
13. POWE 33/353 Clarke 'Minute. Petroleum Policy', 21.4.23.
14. *Ibid.*
15. POWE 33/353 'Petroleum – Nationality restrictions on oil leases . . .', 18.7.23.
16. POWE 33/253 Board of Trade (Petroleum Dept.), 'Petroleum Industry in the British Empire', Feb. 1928.
17. POWE 33/253 H.P.W.G. 'British Empire Production. Supplementary Note in reply to points raised by Mr. Grylls', Feb. 1928.
18. In 1928 both colonies increased production to 1.1 million tons and 0.8 million tons respectively. (FO 371/13540 Cole to Craigie, 8.7.29).
19. POWE 33/253 H.P.W.G. 'British Empire Production. Supplementary Note in Reply to points raised by Mr. Grylls', Feb. 1928.
20. Board of Trade, 'Report of National Fuel and Power Committee', *PP* 1928–29, Vol. 6, Cmd. 3201, pp.513-50.
21. Imperial Conference, 1930, 'Appendices to the Summary of Proceedings', *PP* 1930–31 Vol.14, Cmd. 3718, pp.701-972, p.924.
22. Imperial Conference, 1930, 'Summary of Proceedings', *PP* Vol. 14 1930–31, Cmd.3717, pp.569-700, p.641.
23. Committee on Imperial Defence, 'Sub-Committee on Oil from Coal. Report', *PP* 1937–38, Vol. 12, Cmd. 5665, pp.439-512, p.451.
24. *Ibid.*
25. American share in Venezuela's production increased from 0.5 per cent in 1920 to 59.5 per cent in 1939. (cf. U.S. Senate, *American petroleum interests in foreign countries*, Hearings before a special Committee Investigating Petroleum Resources, 79 Cong., 1 Ses., 1946, p.337.
26. Calculated from, Venezuela, Ministry of Mines and Hydrocarbons, *Venezuelan Petroleum Industry. Statistical Data* (Caracas, 1966), Table: Petroleum Industry Trends, 1917-1966, p.1.
27. CAB 50/3/0.B.19 Secret, Committee of Imperial Defence. Oil Fuel Board, 'Oil Supplies to the World', July 1926. In the case of war, Venezuelan crude could be counted on whereas oil from the colonies and the Middle East would be unreliable.
28. Venezuelan crude was also used by Mexican refineries to supplement waning production in that country. (cf. 'Venezuela, Development of the Petroleum Industry', *Board of Trade Journal & Commercial Gazette* (London) CXX: 1629 (23 Feb. 28), 254.
29. Davenport & Cooke, *op. cit.*, p.228.
30. POWE 33/572 'Imports of Crude Petroleum and Refined Products into the United Kingdom during the year 1938'. Undated.
31. Sir Henri Deterding, *An International Oilman* (London: Ivor Nicholson & Watson Ltd., 1934), p.97.
32. cf. B.S. McBeth, 'Juan Vicente Gómez and the Venezuelan Oil Industry, with special relevance to British oil companies', (B. Phil. Thesis, Oxford University, 1975), p.66.
33. Up to 1945 the refineries on these islands accounted for 74 per cent of total South American refined products. (cf. U.S. Senate, *Wartime Petroleum Policy under the Petroleum Administration for War*, Hearings before a Special Committee Investigating Petroleum Resources, 79 Cong., 1 Ses. 1948 (Washington: USGPO, 1948).
34. J.E. Hartshorn, *Politics and World Oil Economics* (New York: Frederick A. Praeger Publishers, 1962), p.108.
35. For example, the industry's response to high prices and shortages during the

World War I resulted in nearly 34,000 new wells being drilled in 1920 (compared to slightly over 14,000 in 1914), a number not exceeded until 1948.

36. Ronald B. Shuman, *The Petroleum Industry. An Economic Survey* (Norman: University of Oklahoma Press, 1940), p.62-3.

37. P.H. Frankel, *Oil, the facts of life* (London: Weidenfeld & Nicolson, 1962), p.19.

38. In 1940 the average cost in the U.S. was $17,300. This figure varied from $1,000 in Kentucky to $42,000 in California. (cf. Shuman, *The Petroleum Industry*, Table 8, pp.54-5).

39. Because of these different market conditions there is no relationship between the short-run price of crude oil and that of the refined product prices. A number of authors have examined retailers' gross margins from crude production to selling stations and have found that, for 1920-52, each price mark-up varied, demonstrating that there is no significant correlation between them. (cf. Ralph Cassady jr., *Price making and price behaviour in the Petroleum Industry* (New Haven: Yale University Press, Petroleum Monograph Series No.1, 1954) Chart 2, p.136 & Table 13, p.137; de Chazeau & Kahn, *Integration and Competition in the Petroleum Industry*, Exhibit IV-1, p.86; U.S. Congress, Investigation of Concentration of Economic Power, Part 15, *Petroleum Industry*, Hearings before the Temporary National Economic Committee (Washington: USGPO, 1940); and, U.S. Tariff Commission, *Petroleum*, War Changes in Industry Series Report No. 17, (Washington: USGPO, 1946).

40. Edward H. Shaffer, *The Oil Import Program of the U.S.* (New York: Frederick A. Praeger Publishers, 1932), p.3. For example, the effect of the discovery in 1930 of the East Texas oilfield was for crude oil prices to drop from $1.19 per barrel in 1930 to $0.65 in 1931 (a 45 per cent drop), while petrol prices for the same period dropped by only 23 per cent. The fall in crude oil prices increased the number of refineries from 346 in January 1931 to 454 in January 1934. This new competition stimulated refiners to change their selling techniques in an effort to protect their market share, with attention given to 'the use of advertising, the development of brand names, the upgrading of the quality of products, and the improvements of service station facilities and services'. (cf. John G. McClean & Robert W. Haigh, *The Growth of the Integrated Oil Companies* (Norwood (Mass.)): Plimpton Press, 1954), p.108.

41. Shaffer, *The Oil Import Program*, p.4.

42. *Ibid.*, p.4.

43. U.S. Congress, *Petroleum Industry*, 1940, Testimony, Christopher Del Sesto, Special Assistant to the Attorney General, Department of Justice, p.9607.

44. Cf. Leonard Bates, *The origins of the Teapot Dome. Progressive Parties and Petroleum, 1909-1921* (Urbana: University of Illinois, 1963); Nash, *The United States Oil Policy;* and, Burl Noggle, *Teapot Dome. Oil and Politics in the 1920's* (New York: W.W. Norton & Co. Inc., 1965).

45. FO 371/12046 Dept. of Trade, British Embassy (W'ton), 'Memorandum on Report by the Federal Oil Conservation Board, 6 September 1926', 25.7.27.

46. U.S. Senate, *Regulating Importation of Petroleum and Related Products,* Hearings before the Committee on Commerce, 71 Cong., 3 Ses. (Washington: USGPO, 1931).

47. U.S. Congress, *Petroleum Industry*, 1940. Statement by Dr. Joseph E. Pogue.

48. Roy C. Cook, *Control of the Petroleum Industry by Major Oil Companies,* Temporary National Economic Committee - Investigation of Concentration of Economic Power - Senate Committee Print Monograph No. 39, 76 Cong., 3 Ses. (Washington: USGPO, 1941).

49. Robert Engler, *The Politics of Oil - A Study of Private Power and Democratic Directions* (New York: The Macmillan Co., 1961), p.141.

50. *Ibid.,* p.149.
51. De Chazeau & Kahn, *Integration and Competition in the Petroleum Industry,* p.164.
52. *Ibid.*
53. The American oil industry was one of the few major industries which emerged out of the economic depression of the thirties in a better financial position. This was because the companies during the years did not invest in new machinery but still had considerable cash flows which they used to reduce their funded debt; for example, 15 of the major oil companies reduced their debt from $820.3 million in 1929 to $675.9 million in 1933, an 18 per cent reduction. (Cf. 'Financial position of American oil industry strengthened during Depression', *Petroleum Press Service,* 1:18 (15 Sept. 34), 4-5).
54. Transfer pricing, for example cf. E.T. Penrose, 'Middle East Oil: The International Distribution of Profits and Income Taxes', *Economica,* 27:107 (Aug.1960), 203-213.
55. Raymond F. Mikesell & Hollis B. Chenery, *Arabian Oil* (Chapel Hill: The University of North Carolina Press, 1949) p.155.
56. M.A. Adelman, *The World Petroleum Market* (Baltimore: The Johns Hopkins University Press, 1972).
57. Vernon Herbert Grigg, 'The International Price Structure of Crude Oil' (Ph.D. Diss., Massachusetts Institute of Technology, 1954), p.49.
58. The situation of a large number of independent crude producers in the U.S. warranted the need to adopt this system. The U.S. Gulf was practically the only place where importers could obtain supplies and spot cargoes on the open market to cover any likely requirements. For any importer in a foreign country, the U.S. always offered an alternative source of supply. As a result a large part of the world's pre-World War I demand was met by U.S. oil exports from the Gulf coast. For non-U.S. sources after the war this system was not needed as most crude was produced by large vertically integrated oil companies without the need for a market price. (Cf. Helmut J. Frank, *Crude Oil Prices in the Middle East – A Study in Oligopolistic Price behaviour* (New York: Frederick A. Praeger Publishers, 1966); and, Walter J. Levy *The Past, Present and Likely Future Price Structure for the International Oil Trade* (Leiden: E.J. Brill, 1951).
59. Cf. Billy Hughel Wilkins, 'The effects on the economy of Venezuela of action by the international petroleum industry and the United States regulating agencies', (Phd. Diss., University of Texas, 1962).
60. J.B. Dirlan, 'The Petroleum Industry' in Walter Adams ed., *The Structure of American Industry. Some case studies* (New York: The Macmillan Co., 1954), 236-273, p.249. In addition the 'official' Gulf oil prices used were the 'open spot' transaction (i.e. prices reported by the refiners, by-product pipeline and tanker terminal operators) published in *Platt's Oilgram Price Service.* These prices did not include discounts offered or purchases on the futures market, inter-refinery transactions. During time of shortages prices were witheld to new customers. Moreover, the bulk of international oil trade moved on long-term contracts, and the prices quoted did not cover this. Thus:

> what Platt's quotations actually cover is a relatively thin market, limited to transactions between major or independent suppliers on the one hand and independent United States marketers, foreign refiners, and importers (not affiliated with any of the majors) on the other. Moreover, during periods of shortage the quotations become even less representative by recording only what some sellers would charge *if* trade conditions were different. Though they probably account for but a minute portion of all refined products moving out of United States Gulf coast ports, these quotations are used to

establish the 'base price' on which price levels over the world are based'
U.S. Federal Trade Commission, *The International Petroleum Cartel*,
p.351.

61. Penrose writes that the basing point system, which is what the Gulf + system
     was, is a very 'effective device for ensuring not only that uniform prices are
     quoted by all sellers but *also that low-cost producers cannot increase their share of
     the market by reducing prices'*. (Edith Penrose, *The International firm in
     developing countries. The International Petroleum Industry* (London: George
     Allen & Unwin, 1968), p.180.
62. George S. Gibb & E.H. Knowlton, *The Resurgent Years, 1911-1927* (New
     York: Harper & Bros., 1956), p.79.
63. U.S. Federal Trade Commission, *The International Petroleum Cartel*, p.199.
64. U.S. Senate, *Patents*, Hearings before the Committee on Patents, 77 Cong., 2
     Ses. (Washington: USGPO, 1942). Testimony of Patrick A. Gibson, Assistant
     Attorney General, Antitrust Division, Department of Justice, p.3337.
65. U.S. Senate, *Foreign Contracts Act*, Joint Hearings before a Sub-Committee of
     the Committee on the Judiciary, U.S. Senate and the Special Committee
     Investigating Petroleum Resources, 79 Cong. 1 Ses. (Washington: USGPO,
     1945), Statement, Wendell Berge, Assistant Attorney General of the U.S.,
     p.48.
66. The first meeting took place on 6 November 1926, at the Plaza Hotel, New
     York. A second meeting was arranged on 2 August 1927 at the Hotel
     Europäischer Hof, in Heidelberg.
67. On 1 April 1928, *I.G. Farben* opened a new factory at Leunawercke to produce
     synthetic oil. Only 2 per cent of Germany's total coal production was needed to
     supply German oil needs. (cf. 'El petróleo sintético', *Boletín de la Cámara de
     Comercio de Caracas*, 17:175 (1 June 28), 41002-3.
68. U.S. Senate, Hearings on Patents, Testimony, Patrick A. Gibson, Special
     Assistant to the Attorney General, Antitrust Division, Dept. of Justice,
     Exhibit 1, 'Memorandum of Meeting, 31 August 1928, 2:00 pm', p.3429.
69. On 21 March 1929, Messrs. Bosch, Gaus, Schmitz, von Knieriem of *I.G.
     Farben*, and Teagle, von Riedemann, Clark, Howard and Haslan of *Exxon* met
     to review the 1928 summer talks and agreed that the companies would remain
     separate. As a result on 9 November 1929, the 'Division of Fields Agreement'
     was signed in which *I.G. Farben* agreed to stay out of the oil business and *Exxon*
     out of the chemical business. *I.G. Farben*, however, would retain the oil
     business in Germany and *Exxon* would trade chemicals in the U.S. only as a
     junior partner of *I.G. Farben*. On the same day the 'Four Party Agreement'
     between *Exxon, Standard-I.G. Farben* (a patent licensing company owned by
     Exxon and *I.G. Farben* but controlled by *Exxon*), *I.G. Co.* and *Standard
     (Delaware)* was signed. By this agreement *I.G. Farben* assigned to *Standard-
     I.G. Farben* the exclusive control of all present and future hydrogenation
     processes, both patented and unpatented, 'but only insofar as these processes
     were useful in the oil industry'. (U.S. Senate Hearings, *Patents*, Testimony
     John R. Jacobs, Special Attorney, Anti-Trust Division, Justice Dept., p.1753)
     Exxon and *Shell* formed the *International Hydro Patents Co.* to receive, hold,
     and license foreign hydrogenation patents. The agreement was a:

> life insurance to Standard against depletion of crude. It meant protection to
> Standard against promotion of synthetic production of I.G. or licenses
> under I.G.'s processes as then existing or as might be thereafter developed.
> It meant protection to Standard against ventures into synthetic production
> in foreign countries, where Standard marketed, by nationalistic govern-
> ments seeking self-sufficiency'. (*Ibid.*, p.3337).

With this agreement the two companies acquired the unrestricted right to licences:

> Licenses were not, however, to be offered to the remainder of the industry if the effect would disturb marketing position, or as it was termed in the agreement, the 'as is position'. This expression used to adapt patents to markets, was defined in detail. In brief it meant the marketing position of respective companies in respective countries as enjoyed in the year 1928. (*Ibid.*, p.3350).

70. The Corporation had 100 shares divided as follows: *Exxon 50, Standard Oil (Louisiana) 25, Humble Oil & Refining Co. 20, & Carter Oil Co. 5.*

71. U.S. Federal Trade Commission, *The International Petroleum Cartel*, p.219.

72. *Ibid.*, p.220.

73. The 16 companies were: *Atlantic Refining Co., Cities Service Co., Continental Oil Co., Gulf Refining Co., Maryland Oil Co., Pure Oil Co., Richfield Oil Co., Shell-Union Oil Corp., Sinclair Consolidated Oil Co., Standard Oil (Louisiana), Standard Oil (California), Standard Oil (New York), The Texas Corp., Tidewater-Associated Oil Co., Union Oil of California,* and *Vacuum Oil Co.*

74. A.C. Veatch, 'Oil, Great Britain and the United States', *Foreign Affairs,* 9:4 (July 1931), 663-73, p.665.

75. U.S. Federal Trade Commission, *The International Petroleum Cartel*, p.271. The other provisions were: (1) the principle of the Pool Division would apply to every country of the world except the U.S.; (2) it was to apply to supply (that is production and exports) and distribution of crude oil and refined products; (3) the supply of Pool Division oil would be handled by a committee in New York; (4) the distribution of such oil would be handled by a committee in London; (5) these two Committees would be co-ordinated by a central committee in London.

76. U.S. House of Representatives, *Production costs of crude petroleum and refined products,* House Doc. 195, 72 Cong. 1 Ses., 1932, Table 25, p.49.

77. *Ibid.*, Table 27, p.53.

78. It should be borne in mind that:

> Even if there should be no change whatever in the character of the domestic crude or the foreign crude compared, the relation of their values would appear different at different times. Thus, the net values of the products from domestic crude was materially lower in 1930 than in 1929 chiefly by reason of lower prices of the several finished products. The net value of the products from Venezuelan crude showed less decline for, although the prices used in the calculation were the same as for domestic crude, the yield of gasoline from the foreign crude was much higher in 1930 than in the year before, chiefly as the result of the installation of cracking equipment in one of the largest refineries handling this crude. (*Ibid.*, pp.53-4).

79. Refiners using other foreign oil made a profit of $0.26 or 18 per cent per barrel of oil. It should be noted, however, that the price collapse in the U.S. oil industry was not felt uniformly throughout the industry.

80. U.S. House of Representatives, *Effects of Foreign Oil Importers on Independent Domestic Producers,* A report of the Subcommittee on oil imports to the Select Committee on Small Business, *House Report No. 2344,* 81 Cong. 2 Ses. (Washington: USGPO, 1950). Testimony of S.A. Swensrud jr.

81. U.S. Tariff Commission, *Petroleum,* War Changes in Industry Series No. 17 (Washington: USGPO, 1946), p.75.

82. U.S. Federal Trade Commission, *The International petroleum Cartel*, p.272.

83. The agreement came to an end at the outbreak of war in September 1939.

BOP-E*

84. U.S Federal Trade Commission, *The International Petroleum Cartel*, p.163.
85. Hereinafter *SOV*.
86. Hereinafter *MGO*.
87. U.S. Federal Trade Commission, *The International Petroleum Cartel*, p.163.
88. *Ibid.*, p.170.
89. *Ibid.*, p.170.
90. This amounted to over 3 million acres in exploitation and over 400,000 acres of exploration concessions.
91. U.S. Federal Trade Commission, *The International Petroleum Cartel*, p.172.
92. *Ibid.*, p.172.
93. *Ibid.*, p.172-3.
94. *Ibid.*, p.174.
95. *Ibid.*, p.182. Production was to be of total recoverable oil, and the ratio of production was fixed at 1 barrel for *MGO* to 3.45 barrels for the *Lago Pet. Corp.* The control of oil was different for in this case *Exxon* received *Lago's* full production quota, which was 50 per cent of *MGO*'s production. In 1938 this was reduced by half when *Shell* became a partner of *Exxon*. Nevertheless, by 1938 for every barrel of oil received by *Gulf Oil*, *Exxon* received 7.4 barrels. The agreement was cancelled on 13 January 1943.

Britain Remains Dependent on Foreign Oil

During the 1930's the international price structure for oil which had been erected since the 1928 Pool Division Agreement ensured the stability of petrol prices in the United Kingdom.[1] Although by then Venezuela had become the largest single supplier of British oil needs, Britain's dependence on *Exxon* and *Shell*, though somewhat diminished by *Anglo-Persian*, was still great. In addition the U.S. and Anglo-Dutch companies controlled Venezuela's output, and U.S. oil supplies still accounted for a large amount of Britain's needs. The advantage which the U.S. held over Venezuela was that it was politically stable. However, the 'closed door' policy pursued by the British was a source of hindrance to British oil companies operating in the U.S., as they were subject to constraints in bidding for Federal oil leases by the application of the 1920 Mineral Oil Leasing Act.

Under this law, federal lands could not be leased to a company registered in what was termed as a non-reciprocating country such as Britain. On 12 February 1929, in a letter to the Commissioner of the General Land Office of California, E.C. Finney, First Assistant Secretary of the Department of Interior, wrote that his department refused to approve the assignment of an oil and gas prospecting permit to the *Shell-Union Oil Co.* (a *Shell* subsidiary) on the grounds that while Holland was recognised as a reciprocating country, it was 'not known that any other country whose citizens are stock holders in the companies involved has been so recognised'.[2] Moreover, the U.S. had received information that *Shell* was controlled by the British government. The company furnished Finney with evidence that it was not controlled by British interests but by Dutch citizens. This attitude was a continual source of anxiety as the U.S. was still a principal supplier of Britain's fuel oil requirements. For this reason alone, Agnew of *Shell* felt that

all restrictions within the British Empire should be removed as quickly as possible. On the contrary, bitter experience has shown us that these restrictions constitute a continuing menace to the security of an important part of our property in the United States and I would again urge in the strongest manner that everything possible should be done to remove this restriction at an early date.[3]

At the same time the Oil Fuel Board,[4] a Sub-Committee of the Committee of Imperial Defence, set up in March 1925 was examining the oil requirements needed by the UK in the event of war, assuming that the U.S. was either 'friendly' (i.e. it supplied oil for civilian use) or 'unfriendly' neutral (i.e. it did not supply oil for civilian use – an event which would call for drastic rationing). The Oil Fuel Board found that if the U.S. was an 'unfriendly' neutral then Britain would be forced to rely on Venezuelan crude at the rate of 1.75 million tons of crude and 3.75 millions of tons of refined products per annum.[5] In the case of the U.S. being an 'unfriendly' neutral then Venezuela, as Table XII shows, would supply 39.3 per cent of the military needs (east and west of Suez), and 44.1 per cent of the civilian needs of the UK. The Empire's fuel requirements in the case of the U.S. being 'friendly' would be met from the sources outlined in Table XIII.

The Sub-Committee's report concluded that the dependence of the Empire on the U.S.A. and Venezuela for 'oil and refined products in time of war gives cause for great uneasiness',[6] and in the event of the U.S. being 'unfriendly', the 'dependence on Venezuela becomes alarming'.[7] It recommended the drilling of new wells in Venezuela and the expansion of the Curaçao refinery, as well as stimulating the development of refineries in Britain. But production in Venezuela, as we have seen, was controlled by two companies neither of which was British. By 1932 production in the British Empire and the Persian oilfields was a meagre 5 per cent of the world's total oil output.[8]

The Petroleum Department, which consisted of three permanent members and functioned as an advisory committee to the government, now sought to press for an 'open door' policy because it felt that British commercial interests would be harmed if other countries, notably the U.S., adopted similar British insular policies. The arguments for retaining the 'closed door' policy were that it was more desirable on general grounds that the Empire's oilfields should be developed by British capital; that it was necessary to take precautions not only to protect the oilfields in time of war, but also

TABLE XII

United States of America – Unfriendly neutral – Suggested sources
from which requirements could be met. (000s tons)

	Crude, Fuel & Gas Oil		Petrol	
	Quantity	Where obtained	Quantity	Where obtained
MILITARY:				
West of Suez	2,377	Venezuela & Dutch West Indies	163	UK (Persia)
East of Suez	1,706	Abadan (Persia)	222	Abadan
	1,008	UK (Persia)		
	750	Trinidad		
	487	Romania		
	675	UK (Venezuela)		
	144	Suez (Persia)		
	231	Rangoon (Burma)		
UK (CIVIL)				
	890	Mexico	187	UK (Persia)
	703	Venezuela & Dutch West Indies	317	Romania
	407	UK (Venezuela)	175	Venezuela & Dutch West Indies
			121	Abadan
			70	Trinidad
			42	Suez (Persia)

Source: CAB 50/3/O.B. 27 Committee of Imperial Defence. Oil Board. 'Sub-
Committee's Report on Oil Supply in Time of War', 20.3.29, Statement E,
Appendix 5, Adapted.

during the period when war was imminent (control of the financial
policy of management was the best insurance against sabotage); that
British oil companies would purchase British equipment whereas
American companies would buy American; that British oil
technical know-how would increase; and, finally, that the experience in India and Trinidad justified the policy. The arguments
against the 'closed door' policy were that the likelihood of finding
large oilfields in the Empire was thought to be poor (if foreign oil
companies found oil so much the better); that foreign companies
were reluctant to hold a minority interest in British companies; that
the exclusion policy prevented foreign capital from assisting in the
search for oil in the Empire; that the 'closed door' policy had created
serious difficulties with the U.S. and could injure British commercial and oil interests in South America where U.S. influence

TABLE XIII

Particulars of unsatisfied requirements after allowing for products obtainable from Persian crude, showing proposed sources from which they would be obtained. (000s tons)

	Crude, Fuel & Gas Oil		Petrol	
	Quantity	Where obtained	Quantity	Where obtained
WEST OF SUEZ				
Military	2,377	USA	163	USA
UK (Civilian)	2,675	USA, Venezuela, Mexico	1,596	USA, Venezuela, Mexico
Canada & South Africa (Civilian)	2,775	USA, Venezuela, Mexico	358	USA, Venezuela, Mexico
Other colonies	22	Romania, Mexico, Venezuela	20	Trinidad, USA
EAST OF SUEZ Military &				
India (Civilian) Ceylon, " Malaya, " Australia (Civilian)	1,912	Burma, Trinidad, USA, Venezuela		
New Zealand (Civilian)	79	USA, Peru	83	USA, Peru
Other colonies (civilian)	22	Romania		
TOTAL	9,862		2,216	

Source: CAB 50/3/0.B.27 Committee of Imperial Defence, Oil Board, 'Sub-Committee's Report on Oil Supply in Time of War', 20.3.29, Appendix 4, Statement B, Adapted.

was powerful; that British companies were now powerful enough to resist any American competition; that undiscovered oil supplies were useless to all; and, finally, that fear of sabotage was ill-founded because people do not readily destroy their own property.[9] The India Office, however, did not feel that Indian capitalists were able to compete on equal terms with foreign capital

in the exploitation of oil bearing lands for the reason that Indian business organisation is in a backward stage of development as compared with America and other foreign countries. Accordingly in the next few years the entry into the field of foreigners with large resources might prove detrimental to Indian national interests since it might have the effect of leaving nothing for Indian capital to exploit when it is in a position to compete.[10]

A remarkable lack of awareness was demonstrated here for, since 1926, large oil companies had combined to divide up the world's oil markets. In India, for example, *Shell* together with *Burmah* formed the *Burmah-Shell Oil Storage and Distributing Co.*, for marketing and distributing oil in the subcontinent. *Shell* and *Anglo-Persian* worked together in the near East and Africa through a joint company, *The Consolidated Petroleum Co.*, for the distribution of oil in South and East Africa, Egypt, Sudan, and Palestine. In the UK, the two companies had also united to form a joint company to store, transport and sell their products. Moreover, although *Shell* was only 40 per cent British owned, it was the policy 'of the British Government to support *Shell* in any matters affecting their trading activities in other countries'.[11]

It was agreed in principle that the pursuance of an 'open door' policy was desirable but that 'oil should continue to be produced as far as possible by British enterprise and that defence considerations should be fully safeguarded'.[12] A new policy would be needed to ensure that a 'proportion of the oil obtained is refined on British territory so as (a) to produce a fuel oil which would be suitable to the Admiralty, and (b) to meet the needs of the War Office and Air Ministry for petroleum spirits'.[13] The Petroleum Department thought that this could be guaranteed whether the companies were British owned or not. As a result the policy in operation should be changed in order to allow foreign companies to function within the Empire on the following conditions: two years' notice of a change in British policy would be given so that British capital would have a final chance of securing concessions; the terms of the existing leases would be maintained and their transfer made subject to government approval; the new policy would be governed on the basis of reciprocity (that is, foreign capital would only be admitted from countries which offered similar advantages to British capital); companies would have to be registered in British territory; a majority of employees and some directors would be British subjects; 50 per cent of the oil obtained would be refined in British territory, and the refinery had to be capable of producing fuel oil suitable for the Admiralty; the government would retain the right of pre-emption in case of emergency; and clauses requiring the concession to be worked in full would be retained. The Petroleum Department felt that these changes would encourage the rapid development of the Empire's oil supplies, and would safeguard in the fullest possible measure the interests of the armed services for supplies of oil in time of war. In 1930 the Cabinet on the recommendation of the Committee of Imperial Defence had already approved in principle this new policy. After consultation had taken place with the companies concerned some of the proposals were put into effect by enacting the proper legislation.

The Petroleum (Production) Act of 1934 vested in the Crown the petroleum rights in Great Britain and gave the British government powers to regulate and issue licences to explore and produce oil. Licences would be issued by the Board of Trade to people who wanted to drill and a levy raised on the amount of oil produced. The licence holder would make his own arrangement with the land-owners for the necessary facilities, but in the event of not obtaining them by agreement 'provision will be made on lines similar to those contained in the Mines (Working Facilities) Acts for securing the grant of the facilities considered by the court to be necessary on terms to be determined by the court'.[14] The Petroleum (Production) Regulations Act made by the Board of Trade on 15 May 1935, under section 6 (1) of the Petroleum (Production) Act, and which came into force on 17 June 1935, permitted the application for licences from foreigners and from companies incorporated outside the UK, provided that an operating company was formed and registered in the UK. A company could apply for a prospecting licence of three years (with an extension of two years) over an area of 8-200 square miles, or mining licences for 50 years (renewable for a further 25 years) over an area of 4-100 square miles. The regulations further provided that the licensee could not assign or attempt to assign his rights to any person other than a British subject or a company incorporated in the UK. If such a company were formed in the UK at least the majority of the directors had to be British. Finally, licences would only be granted to 'reciprocating' countries. Once this legislation was enacted, the U.S. government recognised Britain as a 'reciprocating' country under the 1920 Mineral Oil Leasing Act. The Cabinet also issued instructions to non-self-governing colonies to prepare revised regulations regarding the admission of foreigners and foreign controlled companies on the basis of the British regulations. During the next five years up to the outbreak of World War II 98 prospecting licences were issued, with

16 test boreholes drilled.[15] Exploration was concentrated on South England, South Pennines, Midlands, North East Yorkshire and Midlothians. A small oilfield was discovered in Formby, Lancashire, in 1939, and in June of the same year a larger field was discovered near the village of Eakring in Nottinghamshire. American interests held a total of 32 licences.[16]

The enactment of these regulations was a victory for the U.S. 'open door' policy. It was also recognition that the 'closed door' policy pursued by the British government had failed to develop adequate sources of oil by which the Empire's independence from foreign oil and foreign oil companies might be secured. It is ironic that such a rich and vast Empire as the British one should be devoid of large oilfields. The regulations brought to an end the acrimonious dispute which had been simmering since the end of World War I between the U.S. and the UK over oil supplies, with Britain having to acknowledge U.S. supremacy in oil matters. This recognition was to have important bearing in terms of Britain's commercial relations with Latin America, more especially with Venezuela. It was understood that Latin America was part of America's sphere of influence, with the implication that the U.S. could influence the Venezuelan government against British interests in the country. This was of great significance, since Venezuelan crude and refined products now accounted for nearly 40 per cent of Britain's needs,[17] as we can see from Table XIV given below

TABLE XIV

Percentage imports of crude petroleum and refined products into the United Kingdom, 1935–38

Country	1935	1936	1937	1938
Venezuela & Dutch West Indies	34.0	38.4	40.0	34.1
Iran	20.6	18.6	19.4	19.6
USA	10.6	10.2	13.0	21.3
Iraq	4.9	5.1	3.8	–
Romania	6.1	7.2	5.0	4.1
Mexico	10.1	6.6	5.2	2.1

Source: Adapted. POWE 33/572 'Imports of Crude Petroleum and Refined Products into the United Kingdom during the year 1938', and CAB 50/6/O.B. 247 Committee of Imperial Defence, Oil Board, 'The Expropriation by the Mexican Government of the Properties of the Oil Companies in Mexico – Memorandum by the Petroleum Department', April 8, 1938. Attached Table 'Great Britain – Principal sources of oil supply. The outputs of crude oil of those countries, their relationship to world output during the years 1935, 1936 and 1937'.

The British regulations were also in the same economic spirit which had become dominant in the U.S.. President Franklin D. Roosevelt felt that the 1930 Hawley-Smoot Tariff Act impeded trade, and therefore based his New Deal foreign economic policy on the enactment in June 1934 of the Reciprocal Trade Agreement on Tariff Amendments Act. The Good Neighbor Policy pursued by the New Dealers 'became the means of extending the net-work of trade agreements throughout Latin America'.[18] The stress was now on the 'gospel of peace with trade'.[19] In official U.S. circles it was felt that if manufactured goods could not cross international frontiers, then armies inevitably would. R.C. Lindsay informed Anthony Eden that

> Equality of treatment for all nationals under the unconditional most-favoured-nation policy means economic peace. Trading in preferences, special bargaining arrangements, arbitrary allotments of quotas, artificial diversions of trade in exchange for special concessions, allotments of markets to favoured sellers, preferences granted in the treatment of exchange control – all these are fertile soil for increasing international strife.[20]

While recognising U.S. supremacy in oil matters, the British regulations also had the practical objective of opening up the Empire to foreign companies in the hope of finding new oilfields especially in the event of future war. The revised projections made by the Committee of Imperial Defence on the sources of oil supplies to meet the requirements of military, civilian and industrial needs in the first year of a Far Eastern war demonstrated Britain's overdependence on the U.S. and Venezuelan crude, with Iran the third major supplier. If the U.S. were 'unfriendly' the position would be 'difficult, particularly in view of the doubtful availability of supplies from the British and Dutch East Indies'.[21] In the event of a war in Europe and the Far East, the petroleum requirements for Empire and Allies in the first year of war would be as shown in Table XV.

The regulations, therefore, provided the necessary machinery to harmonise Anglo-American relations over the irritating and very sensitive matter of oil supplies, enabling foreign companies to search for new oil in the Empire. The number of licences applied for increased dramatically from an average of less than one between 1920–34 to 42 in 1935.[22]

TABLE XV

Sources of oil required for Empire and Allies in the first year of war in Europe and the Far East (Million tons)

Country	Mediterranean closed	Mediterranean open
Iran	7.2	8.9
USA	10.6	9.3
Venezuela & Dutch West Indies	4.3	4.3

Source: CAB 50/7/Secret/O.B.290 Committee of Imperial Defence, Oil Board, Sub-Committee on Petroleum Products Reserves, Eighth Report, 'Sources of supplies of petroleum products to meet the estimated requirements of the services and for industry and civil purposes of the Empire in the First Year (1940) of a major war, simultaneously in Europe and the Far East', Undated.

At the same time there was hope that alternative sources of fuel developed in Britain would assume more importance. As we have seen, the 1928 Finance Act levied a four pence per gallon customs duty on imported motor spirit, which was subsequently increased to eight pence per gallon in order to stimulate Scottish shale oil production, but with disappointing results.[23] More successful was *ICI* which during the mid-twenties had experimented with the Bergius process to produce oil from coal. As a result in 1929 it started to build a pilot plant to treat 10 tons of coal per day, running until the end of 1931. Later discussions took place with the government on the possibility of securing development of a larger plant. The government responded by enacting in 1934 the British Hydrocarbons Oil Production Act which would provide for a preference of no less than four pence per gallon for the next nine years in respect of light hydrocarbon oils manufactured in the UK from coal, shale and/or peat indigenous to UK or from products produced from these substances. *ICI* therefore decided to proceed 'at once with the erection of a large scale commercial plant'[24], which started the treatment of creosote in February 1935. The plant had a capacity of 150,000 tons per annum, and produced in 1936 120,000 tons. The British Hydrocarbons Oil Production Act in addition to helping to establish *ICI*'s hydrogenation plant at Billingham had the effect of increasing 'home production of motor spirit from the by-product industry, while the shale oil industry has probably been saved from extinction'.[25]

The British government also hoped for *Anglo-Persian* to increase its participation in Venezuelan oil developments. As we have seen, three companies controlled almost the entire Venezuelan oil production. A number of smaller companies, however, by the late 1920s also produced crude oil. *Anglo-Persian* for example through the *Tocuyo Oilfields of Venezuela* and the *North Venezuelan Petroleum Co. Ltd.* produced in 1932 0.17 per cent of Venezuela's production.[26] In August 1925 Commander Thompson was appointed *Anglo-Persian*'s representative in Venezuela and adjacent States,[27] and given the task of acquiring concessions for the company. However, he encountered opposition from Pedro M. Arcaya, the Interior Minister,[28] who refused to support the desire of his Cabinet colleagues to grant oil concessions to a company 51 per cent owned by the British government. In February 1926, the company came to the conclusion that

> it would be dangerous for the company to risk considerable sums on concessions and their development in Venezuela if the company's legal position were in any way imperfectly defined or secured, since occasions for friction might arise or be created at any time after large expenditure had been made. A merely benevolent attitude on the part of the Venezuelan Government . . . would be of little or no value[29]

In 1929, Reyes Delboso, a Venezuelan, travelled to London together with Jacques Krassny to negotiate Venezuelan oil concessions. But *Anglo-Persian* were interested only if the law were changed to allow them full participation.[30] In March 1934, Sir John Cadman tried to get the Foreign Office to place pressure on the Venezuelan government to change the law. Sir John wrote to Sir Robert Vansittart at the Foreign Office to the effect that Venezuela was so important 'actually and potentially, as a petroleum producing state, that the *Anglo Persian* remains still extremely desirous of getting the Law amended in such a way as to permit it to operate'.[31] He also wrote to Diógenes Escalante, Venezuelan Minister at London, urging him to get the law modified slightly, arguing that the British government's stake in *Anglo-Persian* was practically the same as that of a shareholder who held no more than 50 per cent of the shares, and that the British government had never used its power to veto.[32] *Anglo-Persian* tried again, sending Arthur Charles Hearn,[33] a Director of the company, to Venezuela to get the law changed.[34] Hearn tried unsuccessfully to reach agreement with the *Orinoco Oil Co.* (75 per cent owned by the *Pure Oil Co.*), because his proposals were based on the possibility of making the suitable change in the Venezuelan law. Pedro Rafael Tinoco, Interior

Minister, promised to remove the offending clause but E. Keeling, British Chargé d'Affaires, considered this highly doubtful as Dr. Tinoco was 'supposed to be in the hands of the Standard Oil Company'.[35] Nevertheless, Dr. Zuloaga, *Anglo-Persian*'s legal counsel in Caracas, drafted an amendment to be inserted into the new oil law, which Dr. Tinoco submitted for the consideration of the Cabinet. Cayama Martínez, Development Minister, raised the objections to *Anglo-Persian*'s entrace, considering that public opinion would accuse the government of being bought by the British company.[36] Hearn's efforts were unsuccessful and he was forced to return to London empty-handed.

However, *Anglo-Persian* made better progress in the traditional areas of British interest of the Middle East and the Arabian peninsula. The company with its geologists had first arrived in Qatar in 1924. Later, in 1932, acting as nominees of *IPC* under the Red Line Agreement, the company obtained from the Sheikh a two-year exploration concession. *Exxon* was also interested in securing a concession there, but Qatar was in the British sphere of interest. In 1934, Sir Samuel Hoare, Secretary of State for India, during a meeting of the Committee of Imperial Defence, stated that in his view it

> was absolutely necessary to give the Sheikh an assurance of our protection. He regarded it as an axiom of Imperial policy to prevent intervention by foreign Powers along the southern shore of the Persian Gulf, along which our Imperial air route now ran. He was convinced that the sole measure which would avoid the grant of a concession for oil in the Qatar peninsula to an American company – the Standard Oil Company – was an offer to the Sheikh to guarantee protection against landward aggression.[37]

Lord Hailsham, Secretary of State for War, fearful of further angering the U.S. over the exclusion of American companies from Qatar, said at the same meeting that 'he was doubtful whether the importance of securing the concession to a company registered in Great Britain justified the risks involved'.[38] Nevertheless, the Committee recommended that, in principle, it was desirable that any concessions granted in Qatar should be retained by British companies. On 17 May 1935, *Anglo-Persian* obtained a 75-year oil concession covering 4,100 square miles, having to pay a signature bonus of 400,000 rupees, with an annual rental of 150,000 rupees for the first year and 300,000 rupees thereafter, and a royalty of 3 rupees per ton. The *IPC* formed the *Petrol Development (Qatar) Co. Ltd.*, which on 5 February 1937, took over from *Anglo-Persian*. In October 1938, drilling began and a year later, in December 1939, the Durkan oil field was discovered. In 1935, *Anglo-Persian* together with *Gulf Oil* acquired from Kuwait a 6,000 square mile concession which ran for 75 years. The two companies formed the *Kuwait Oil Co.* to exploit the concession, paying a signature bonus of 470,000 rupees when the contract was signed on 23 December 1934. It would also pay an annual rent of 95,000 rupees, and a royalty of 3 rupees per ton produced, with a minimum provision of 250,000 rupees per annum.[39] The company was not so successful in securing oil concessions in other countries of the Arabian peninsula. In 1933 *Standard Oil Co. (California)* acquired a 60-year concession in Saudi Arabia covering 36,000 square miles,[40] forming the *Californian Arabian Standard Oil Co.* to exploit the concession. The following year another subsidiary, the *Bahrain Petroleum Co.*, obtained an exclusive concession in Bahrain for 55 years covering 100,000 acres. There was much anxiety among the members of the international petroleum cartel that *Standard Oil Co. (California)'s* oil would flood the European market. Initial development work would stem the tide for the time being,[41] but real hindrance lay in the company's lack of marketing facilities. As a result on 1 July 1936, it signed an agreement with the *Texas Corp.* in which the latter company acquired a 50 per cent stake in the *Bahrain Petroleum Co.* in exchange for 50 per cent of the *Texas Corp.'s* marketing facilities east of Suez. A similar agreement was arrived at in 1939 with the *Californian Arabian Standard Oil Co.* The joint marketing company was called *Caltex*, and as the *Texas Corp.* was a member of the oil cartel, the importance of this agreement was that the cartel could now control the rate of development of the oil supplies from Bahrain and Saudi Arabia to its benefit. Nevertheless, despite the discovery of large Arabian oilfields in the mid-thirties, Venezuela remained Britain's largest oil supplier.

On 17 December 1935, Gómez, the strong man of Venezuela who had ruled the country for 27 years, died in his sleep in Maracay. It suddenly became possible to protest against 27 years of political oppression, with the result that on 20 and 21 December mobs in Caracas proceeded to the 'methodical destruction of the houses and property of members of the Gómez family in Caracas, of their associates, and of General Velasco'.[42] In the Zulia oil region the situation was just as bad. In Lagunillas the oilworkers went on the rampage and together with the peons killed as many policemen as they could find, becoming a 'bloodthirsty crowd armed with machetes and they also had about 12 rifles with them'[43], in the words of an eyewitness. On 21 December, a hundred persons were killed and twenty-eight shops were pillaged by the mob, at an estimated loss of £4 million. By Christmas Eve, the situation appeared calmer in Maracaibo but the oilworkers were in a militant mood. According to an official of a *Shell* subsidiary:

> At Lagunillas and Cabimas oilfields the peons went on strike demanding higher wages and other concessions . . . Many peons were killed. One watchman, who had incurred the displeasure of his fellow workmen, was soaked in petrol and set on fire. The strike was finally ended on 27 December after fresh troops had been sent to Maracaibo and from there to the oilfields. The number killed in Lagunillas is put at one hundred and fifty and at Cabimas at fifty.[44]

The strikers did not achieve their demands for higher pay, but did win the introduction of an eight-hour day, medical pay for dependents of employees, improved living conditions, and a change in shifts every two weeks on tour work. The general mood of the country remained, however, antagonistic to the companies. Colonel Seagrin, British Vice-Consul at Maracaibo, reported in April 1936 that:

> There appears to be a good deal of dissatisfaction still among the oilfields' peons and these are becoming more and more difficult to deal with. They are still ill-disposed towards the 'American' – everybody is an American which (sic) is not a Venezuelan – and acts of violence are more common than before. Only a few weeks ago, a German foreman, who had been asked for work by a couple of peons was stabbed to death for the only apparent reason that he had no work to give.[45]

This situation would not be solved until later that year with the general strike of the oil workers.

Following Gómez's death, General Eleazar López Contreras, Defence Minister under the former President, was elected by Congress to head the new government. For political expediency his political administration had to follow the militant mood of the country. The tendency was to develop an anti-foreign feeling, particularly against the oil companies which were accused of exploiting the country's natural resources in collusion with the previous administration. In January 1936, it was reported that the government intended to request Bs.45 million from *Exxon* 'paid as bribes to General Gómez's Ministers'.[46] It was alleged that the company in order to obtain their concessions had paid Dr. Tinoco, the former Interior Minister, a bribe of Bs.20 million, and that if the company did not agree to pay the sum demanded the government would declare all their concessions to have been illegally obtained and cancel them. Moreover, in August the Venezuelan government initiated legal action against *Exxon* and *Gulf Oil* to force them to pay US$14.4 million which the government alleged the companies owed in back taxes for 1927–31.[47] Thus by mid-1936, the companies were seriously worried about the future of their investments, even fearing another Mexico.

British interests were the most affected by this nationalistic tendency as Venezuela occupied such an important position in the UK oil market. Consequently, the problem became for the British government 'a question of national interest as well as the interest of British oil companies which is concerned in any change of policy of the Venezuelan Government affecting the development of the oil industry'.[48] The Foreign Office requested the Petroleum Department to prepare a memorandum detailing the arguments to be used in the event of the nationalisation of the oil industry by the Venezuelan government. The Petroleum Department advised that British capital had contributed substantially to the development of the country's oil resources, that development had been encouraged by reasonable terms for oil concessions, that any changes in these terms would bring about a situation similar to Mexico's, and finally simple blackmail. It was felt that as the UK was Venezuela's largest market for its oil then this position would 'quickly change were conditions to alter so as to make it less attractive for British oil companies to pursue their operations in Venezuela.[49] But as Keeling, British Chargé d'Affaires, had emphasised in 1934 the oil companies 'are so powerful and know the country so well and the various ways of dealing with the people, both open and underhand, that they will get on well under any regime'.[50] The companies were therefore relieved when the new oil law was enacted in August for it presented them with very few problems.[51] Although the oil industry had undergone certain changes during the transitional period between the demise of Gómez and the assumption of power by López Contreras, they were of little importance to the oil companies on the whole, as the new government had 'realised that foreign capital as well as foreign technique'[52] had played a large part 'in the development of the country'.[53] By the autumn of 1936, the indications were that the 'American companies are starting the same somewhat mad scramble for concessions in Venezuela'[54] as they had done in Mexico. For example, *Exxon* took up large concessions in Eastern Venezuela and was also committed to building a large refinery in the small fishing port of Puerto La Cruz, Anzóategui State. Despite this optimistic mood the companies faced considerable problems. On 11 December 1936, the *Unión Sindical Petrolera de Venezuela* called a general strike of all the oil workers in the industry.[55] The strike was so successful that it brought the industry to a standstill, and was only ended by Presidential decree, 43 days after it started, on 23 January 1937. The workers gained an extra Bs.1 on their wage packets.

On 28 July 1937, the Attorney General filed a suit at the Federal Court of Cassation against the *Mene Grande Oil Co.* and *Gulf Oil*, claiming that the companies owed the government Bs.26.6 million in back taxes for 1925–36.[56] The government contended that the companies had paid their royalty tax at 11.25 per cent instead of 15 per cent, thus owing Bs.15.6 million, which with interest payments of Bs.11 million, brought the sum claimed to Bs.26.6 million. On 27 January 1937, *Exxon* in the other tax claim agreed to settle out of court, paying the government Bs.4 million. Four years later, *Gulf Oil* also settled out of court paying the government Bs.30 million.[57]

While these events were occurring in Venezuela, the oil companies in Mexico appeared to be bent on a head-on collision with the Cárdenas government. In 1937, the Mexican Labour Board ordered the oil companies to grant a large wage increase over the minimum 3.5 pesos a day paid to their workers, totalling 26 million pesos. The companies' best offer of 22 million pesos made in early March was not good enough, with the result that on 18 March 1938, Cárdenas nationalised the industry. This posed a significant threat to the British government which was on the brink of entering the greatest conflict in history. The threat was not the curtailment of Mexican supplies (though these would be welcome during the war), but that Mexico's action might trigger off similar behaviour in other Latin American countries, especially in Venezuela. The Committee of Imperial Defence stated that:

> Of much greater importance than the possible loss of Mexican oil in an emergency is the effect which the adoption of a similar policy by other South American oil producing countries – Venezuela, Colombia and Peru – would have. It is, therefore, essential that every effort should be made to ensure that the Mexican policy is not followed by the countries mentioned.[58]

and it was reported by the British Minister at Caracas that the

> Mexican Government has already been making efforts in certain other Latin American countries to secure the adoption of a similar policy. If these efforts were successful a very serious situation would arise. It is unfortunately the case that the Governments of some of these countries are somewhat unstable, while at the moment HM Ministers at Venezuela, Colombia and Peru, who have been asked to report any signs of Mexican efforts do not consider that there is any immediate danger of these countries following suit, the possibility of a change of Government having undesirable consequences is not ruled out by them.[59]

Moreover, in the event of war, U.S. neutrality legislation barred supplies to Britain and her Allies, except Canada. Russian oil exports would cease, and the Dutch East Indies, Romania and Iraq, because of their geographical situation, would be doubtful suppliers. This left Iran, Venezuela and Mexico, together with certain smaller producers such as Colombia and Peru, to keep the British Empire supplied with oil. Furthermore, the Latin American countries were favourably situated from a 'sea transport point of view'.[60] It was apparent, therefore, that if other countries were to follow Mexico's action, British oil supplies would be in serious jeopardy and at the mercy of the Latin American countries. The Foreign Office consequently made strenuous efforts to avoid such a situation developing.

The British Ambassador at Washington approached the State Department to see what could be done. At first, U.S. opinion was difficult to gauge because there were 'certain suspicions that they may not object to the removal of foreign interests in the Latin American countries which in the end may open up a possibility of the United States getting back to the exclusion of the rest'.[61] The officials of the State Department felt that another oil crisis in Latin America would 'jeopardise all United States interests';[62] moreover, if Venezuela were to follow Mexico's lead and nationalize her own industry with adequate compensation (the companies demanded US$600 million), 'the proper interpretation of the Monroe Doctrine will become the gravest problem the State Department will have to face'.[63] Accordingly, the State Department later expressed their understanding of Britain's position to the British Ambassador, with the result that 'in spite of their good-neighbour policy, they will not be surprised that this country [Britain] should desire to take every practicable step to safeguard its position in regard to oil supplies in these Latin American countries'.[64] The Petroleum Department consequently advised that economic pressure be placed on Mexico. An increase in the tariff on Mexican goods would act as a deterrent to the 'other South American oil producing countries, which otherwise might be encouraged to follow Mexico's example'.[65] The Petroleum Department concluded that

> this incident does draw attention to the desirability of watching very closely in future the situation in those oil producing countries on which this country is bound to rely for supplies. It is suggested that special steps be taken to deal with any cause of friction which may now exist or arise in future, that possibly more frequent visits should be paid by HM ships with the object of cultivating good relationships and that HM Ministers in the countries concerned should be asked to advise whether there are any other directions in which they consider that action might be taken which would strengthen our position.[66]

Earl De La Warr, President of the Committee of Imperial Defence (and Lord Privy Seal), taking note of the Oil Board's Memorandum, and realising that Britain 'would be faced with a grave situation if these countries followed Mexico's example',[67] endorsed the proposals of the Petroleum Department to adopt special steps to ameliorate any friction between Britain and the Latin American oil producing countries, and cultivate good relations. A few days later, at the 322nd meeting of the Imperial Defence Committee, the Mexican oil nationalisation was discussed. Lord Swinton (formerly Sir Philip Lloyd Greame), Secretary for the Air Ministry, 'attached the greatest importance to oil supplies from Venezuela'.[68] Sir Robert Vansittart, representing the Foreign Office, agreed that 'retaliatory action should be taken to deter other Latin American countries, more especially Venezuela, from following Mexico's example, although he doubted whether such action would have any appreciable effect'.[69] As a first step it was decided to send HMS *York* on a goodwill trip to Colombia and Venezuela in the summer months. In the ensuing months the British government approached the banks and finance houses to ask them to refuse credits to those who might attempt to trade in Mexican oil. Neville Chamberlain's government deprecated the handling of Mexican oil and refused to allow it to form part of government contracts. Similar instructions were given to colonies and dominions. In addition, British traders were advised not to supply equipment or materials to the Mexican national oil company. The *Mexican Eagle Co.,* a *Shell* subsidiary, used its influence to prevent both the sale of Mexican oil and the supply of equipment to the Mexican national oil company. The company also took legal action to prevent Mexican oil landing in certain countries. For its part the U.S. government withdrew from the silver agreement it had with Mexico,[70] and also took steps to block Mexican oil sales to the Navy.[71] The position was very delicate for if Mexico was pushed too far this might increase nationalistic feelings in the rest of Latin America, while at the same time Mexico's deteriorating economic position would make her appeal to Germany and Japan for help. Nevertheless, these measures met with considerable success, for up to January 1939 no Mexican oil had been imported into the UK.[72]

Events in Venezuela, however, were more influenced by the legacy left by the Gómez dictatorship than by what occurred in Mexico. It should be borne in mind that Mexico's production was not as large as that of Venezuela. Mexico's action did, however, strengthen Venezuela's position internationally because neither the UK nor the U.S. wanted a repeat of the Mexican fracas. Moreover, the oil companies did not have a suitable alternative to replace Venezuelan crude now that Mexico had nationalised its oil industry. It became apparent, therefore, that the future Allies and oil companies would depend even more on Venezuela. There was deep-felt belief in the Venezuelan government that the oil companies in years past had been granted too many privileges, resulting in huge profits without any contribution to the country's welfare. Dr. Nestor Luis Pérez and later Dr. Manuel R. Egaña, Development Ministers during López Contreras' government, campaigned vigorously to achieve a more equitable distribution of profits. After visiting the country in June 1938, Sir Frederick Godber, Chairman of *Shell,* stated that:

> While the Venezuelan Government, it is believed, hold the view that expropriation is not a practical policy, the conviction that the Nation is not getting an adequate participation in the exploitation of its Natural wealth will result in continually increasing demands and burdensome legislation. Moreover, the possibility that an attempt will be made to abrogate the concessions granted in earlier days on more favourable terms cannot be put out of the question. At

the moment it is not thought likely that any such drastic action will be attempted. But the degree of success of other countries in a policy of confiscation and repudiation must be expected to have a definite influence on the future policy of the Venezuelan government in this connection.[73]

Shortly after he took up his appointment as Development Minister in 1936, Dr Pérez started to develop a policy of restricting the rights of the oil companies to customs exoneration. The oil companies had always imported a great deal of their equipment and import duties had in the past been waived on much of it. Dr. Pérez argued that exoneration from customs duties should not be granted on any materials which were produced or manufactured in the country. The companies took the matter up with the Minister and protested vigorously against the curtailment of their privileges. After the refusal by the Minister to consider their point of view the companies took the matter to the Federal Court of Cassation. In April 1938, the Court heard the first case brought by *Exxon,* and ruled in favour of the company,[74] as it was to do in the other cases brought by oil companies. On 13 July 1938, only four months after the Mexican nationalisation, the Venezuelan government enacted a new oil law. It gave the government greater royalty payments as well as increasing the surface and exploitation taxes. For the oil companies, the law contained a number of objectionable clauses, as the government would now determine unilaterally the value of royalty cash payments (previously this had been done by negotiation). The government would no longer grant direct exploitation concessions and the exoneration from customs duties was further reduced. The companies would have to submit a list ('lista previa') of items they wanted exempted to the Development Minister for his approval. The law had the effect of restraining companies from taking up new concessions.

Dr. Egaña, who shortly after this replaced Dr. Pérez as Development Minister, refused outright to grant exoneration from customs duties on the grounds that the materials in question were not essential to working the oil concessions. He even applied this policy to those companies which had obtained a favourable ruling from the Court of Cassation. The companies filed suits in the same Court, and pending the outcome of the cases continued to submit the 'lista previa', which the Minister refused to accept. According to the Petroleum Department:

These suits against the Government are regarded as particularly important not only on account of the very large amount of money involved, but because there is little doubt that the issue has ceased to become one personal to the Minister of Fomento (sic), but has now been adopted by the Government as part of their policy of trying to exact from the Oil Industry additional revenue to compensate for what in their minds are the too favourable terms of the earlier concessions.[75]

The Development Minister insisted that while the Court might reverse his decision in specific instances, this did not bind him in respect of other cases, and continued to deny exoneration on a number of articles. A peculiar situation arose when Dr. Parra, Finance Minister, allowed the companies to import their equipment free of import duties, while Dr. Egaña carefully recorded the value of goods entering the country in order to bill the oil companies later.[76] This, together with the Cassation Court's ruling in April 1938 that the *Mene Grande Oil Co.* and *Gulf Oil* owed the government Bs.15.6 million in back taxes,[77] brought relations between the oil companies and the government to a low ebb. The U.S. government, however, wanted to avoid at all costs a repeat of the Mexican experience. President Roosevelt was of the opinion that the Mexican crisis was created by the intransigence of the oil companies,[78] and therefore in a future conflict it was vital to ensure that Venezuelan crude was delivered to the Allies rather than to the Axis powers. Between 1938 and 1939 Germany had obtained 44 per cent of her oil supplies from Venezuela.[79] In the nation's interest it was necessary to get the oil companies to take a more conciliatory attitude. In 1939 the State Department recommended that:

> In order to avoid these difficulties, I believe this government must be prepared to go further than may be customary in advising the American petroleum companies in the course they should pursue. It must not be permitted them (as occurred in the case of the Mexican dispute) to jeopardise our entire Good Neighbor policy through their obstinacy and short-sightedness. Our national interests as a whole far outweigh those of the petroleum companies. I think that it would be proper to inform the President of Venezuela or the Ambassador of Venezuela here that we should like to hear from them at any time of any wrongdoing they believe has been committed by the American companies since we expect our companies to conduct themselves in full accord with the

letter and spirit of the law just as we expect that they will be treated similarly.[80]

Before taking up his post as new U.S. ambassador at Caracas in 1939, Frank P. Corrigan saw the Presidents of the oil companies operating in Venezuela in New York to

> emphasise the Department's desire to cooperate in every practicable way towards resolving difficulties which the oil companies may experience before they assume more important proportions and to emphasise the importance of the oil companies having in Venezuela the best type of representation to maintian a consistently cooperative attitude in relation with the Venezuelan government and people.[81]

In order to underline this new relationship a Reciprocal Trade Agreement was signed with Venezuela on 6 November 1939, a month after war was declared in Europe.[82] Under the agreement, both countries would reduce import tariffs on a number of goods, the most important of which was the reduction by 50 per cent of the U.S. oil tariff for Venezuelan oil. By this agreement, Venezuelan crude was tied firmly to the U.S. market, securing the oil required for the Allies.[83] Moreover, Venezuelan crude was further favoured when on 12 December, a U.S. Presidential Proclamation reduced U.S. import tax on foreign oil by 50 per cent on an annual quota not exceeding 5 per cent of U.S. continental refining capacity, of which Venezuela was allocated 72 per cent.

These developments in Latin America, together with the enactment in the U.S. of the second Neutrality Act,[84] which allowed American oil to be available for war as long as it was paid in dollars and transported by non-U.S. ships, reinforced the need for Britain to secure safer fuel supplies. A renewed interest was shown in the development of techniques to produce oil from coal. On 26 April 1937, Sir T. Inskip, the Minister for the Co-ordination of Defence, appointed a Sub-Committee of the Committee of Imperial Defence (chaired by Lord Falmouth) to enquire into the production of oil from coal. The Sub-Committee looked at the hydrogenation of coal and production of oil by the synthesis of gases, and was quite satisfied that the Department of Scientific and Industrial Research had the technical capacity which would enable both processes to be tried out experimentally.[85] Nevertheless, there were strong economic and strategic reasons against adopting the hydrogenation process. In Germany, for instance, where both the

Fischer-Tropsch and Bergius hydrogenation processes had been developed 'as part of the policy of the German Government to make Germany self-supporting in regard to supplies of motor fuel',[86] the Sub-Committee agreed that at the level of oil prices then prevailing the 'case for home-produced oil, judged by purely economic standards, falls to the ground'.[87] Something like 7 per cent of Britain's motor spirit demand was supplied from home sources, while no fuel or lubricating oil and other products were supplied. Thus, in order to play an important role in war, a number of processing plants at a large capital cost would be required. The extra supplies of coal needed would take time to produce, whereas oil supplies could be turned on almost immediately. It was also felt that oil tankers, because of the large number of them, stood a better chance of getting to Britain during war-time, whereas large hydrogenation plants would be relatively easy to bomb. In addition, the government would have to invest additional funds in marketing and distribution facilities for the oil produced during peace-time from the hydrogenation plants. The Sub-Committee concluded that

> in general a policy depending on imported supplies with adequate storage, is the most reliable and economical means of providing for an emergency, and they cannot recommend the reliance of the country in war time on supplies of oil from indigenous sources especially established for this purpose, unless any particular aspect of the case can be shown to be exceptional.[88]

The Sub-Committee therefore recommended that the preference subsidy on home produced oil be extended for a further 12 years from 1938, and be increased from four pence to eight pence; that the Department of Scientific and Industrial Research continue their investigation into low-temperature carbonisation, and finally, that the Fischer hydrogenation process be tested in commercial production of 20-30,000 tons.

In 1938 85 per cent of the UK's demand for refined products were marketed by *Shell* (40 per cent), *Exxon* (30 per cent), and *Anglo-Persian* (15 per cent). During the summer of 1938 these oil companies together with several smaller ones were invited by the government to draw up their own administrative plan for war. As a result the Petroleum Board, under Sir Andrew Agnew, a *Shell* Director, was created to act as a unified central body which would coordinate supplies and distribution of essential products.

In 1938 the Oil Board felt that Britain could contemplate war with Germany without U.S. oil stocks. It felt that only aviation spirit and lubricating oil would be in short supply, but with the development during the 1930s of a solvent refining process, supplies of lubricating oil could be increased using Middle Eastern and Venezuelan oil to cover all requirements other than for aircraft engines. Special lubricating oil for aircraft engines ('bright stock') needed to be stockpiled because only the U.S. and the USSR produced them. The Air Ministry had already laid in its own reserves to meet the problem, as well as offering to sign long-term contracts with firms that produced the oil, something which was unsuccessful. The Air Ministry was more successful in procuring high octane aviation fuel, signing in 1937 a number of contracts with the *Trinidad Leaseholds Ltd, Exxon* and *Shell,* which together with *Anglo-Persian*'s two plants at the Abadan refinery in Persia meant that the Air Ministry could draw on 850,000 tons of 87 octane petrol a year outside the U.S. This would be an ample supply as it was calculated that only 650,000 tons would be needed for the first year of war. However, it was later realised that aircraft would be running on 100 octane fuel, of which supplies were limited. Under Sir Harold Hartley, the Aviation Fuel Committee of the Air Ministry recommended in December 1938 that the Air Ministry finance the construction of three processing plants to produce 720,000 tons of 100 octane spirit per annum. In January 1939 the Imperial Defence Committee approved this plan and called on *ICI, Shell* and the *Trinidad Leaseholds Ltd.* to form a company to supervise the construction of one plant at Heysham (Lancs.) and two in Trinidad to avoid air attacks.

The Oil Board in January 1939 estimated that fuel demands for the armed services in case of a war in Europe and Japan would be around 10.25 million tons or equivalent to Britain's total consumption needs for 1938. Stockbuilding was slow not because of supply constraints but because of a lack of storage capacity, with the result that at the outbreak of war none of the three services had stockpiled sufficient products for its needs.[89]

Despite the efforts made by the British government to open the Empire to foreign oil companies, and its desire to explore the possibility of producing synthetic fuels from Britain's large coal and shale reserves, the country remained linked to the international oil industry, dependent on foreign oil (now mainly from Venezuela) supplied by *Shell* and *Exxon.* Britain's efforts to secure its own crude oil supplies and thus lessen its dependence on foreign oil companies had failed.

NOTES

1. According to the U.S Federal Trade Commission:

   The long-term cooperation of the national companies Standard (New Jersey); Shell, and Anglo Iranian – and their predominance in the market provided the core of this control. Built around this core, and supported by it, was a scheme of control that may be likened to private licencing of distribution and retailers, although the scheme was less successful in the control of retail trade. The national companies applied the 'as is' principles among themselves, and these principles were, in effect, transmitted to the independent distributors through modified cartel-like arrangements. Except for sporadic rivalry among retailers, resulting largely from the activity of the 'pirates', price competition had been almost completely eliminated from the petroleum industry by the latter thirties. US Federal Trade Commission, *International Petroleum Cartel*, p.320.

2. FO 371/13540 Andrew Agnew *(Anglo-Saxon)* to Cole, 7.5.29.
3. *Ibid.*
4. The Board was set up on the advice of the Principal Supply Officers of the three services and was composed of representatives of the three armed services, the Colonial and Dominions Offices, Mines and Mercantile Marine Departments of the Board of Trade, Department of Scientific and Industrial Research, and the Treasury. The Chairmanship was always under the Admiralty's Civil Lord. (cf. Payton-Smith, *Oil.*)
5. CAB 50/3/Secret O.B.27 Committee of Imperial Defence, Oil Board, 'Sub Committee's Report on Oil Supply in time of War', 20.3.29.
6. *Ibid.*
7. *Ibid.*
8. cf. John R. Bradley, 'Fuel and Power in the British Empire', U.S. Bureau of Foreign and Domestic Commerce, *Trade Promotion Series* No. 161, 1935. Trinidad was the largest single producer in the British Empire.
8a. *Hansard*, 1933–4, Vol. 288, F. Brown, 26.4.34.
9. BT 11/41B/CRT 2241 Petroleum Department, 'Memorandum. Oil Concessions in British Colonies and Protectorates. British Control of Companies', 25.7.30.
10. POWE 33/275 India Office, 'Memorandum on the grant of oil concessions in British India', 1.11.29.
11. POWE 33/461 Petroleum Department, 'International Control of Petroleum', Oct. 1932.
12. BT 11/41B/CRT 2241 Petroleum Department, 'Memorandum. Oil Concessions in British Colonies and Protectorates. British Control of Companies', 25.7.30.
13. *Ibid.*
14. *Hansard*, 1933–4, Vol. 287, Walter Runciman, President of the Board of Trade, 22.3.34, col. 1382.
15. *Hansard*, 1938–9, Vol. 349, G. Lloyd, Secretary for Mines, 11.7.39.
16. Payton-Smith, *Oil*, p.17n.
17. The total is higher if we take into account the unknown amount of refined products exported from the U.S. using bonded Venezuelan crude.
18. Edith James Blendon, 'Venezuela and the United States, 1928–1948: the impact of Venezuelan nationalism' (PhD.Diss., University of Maryland, 1971), p.55.
19. BT 11/591 Des.136E R.C. Lindsay to A. Eden, 5.2.36.

20. *Ibid.*
21. CAB 50/6/O.B.195 Committee of Imperial Defence, Oil Board, Sub-Committee on Petroleum Products Reserves, Fifth Report, 'Sources of Supply of Petroleum and Petroleum Products to meet the estimated requirements of the services and for industry and civilian purposes of the Empire in the first year (1940) of a Far Eastern War', undated.
22. *Hansard,* 1934-5, Vol.303, Capt. Crookshank, 25.6.35.
23. See p.91.
24. Committee of Imperial Defence, 'Sub-Committee on Oil from Coal', p.465.
25. *Ibid.,* p.485.
26. R.J. Kirwin, *Economic Conditions in Venezuela,* Board of Trade, Department of Overseas Series (London: HMSO, 1932).
27. FO 371/10603 Cadman to Foreign Office, 25.8.25.
28. FO 371/10603 *Anglo-Persian* to Foreign Office, 12.11.25.
29. FO 371/11109 *Anglo-Persian* to Foreign Office, 12.2.26.
30. Archivo Histórico de Miraflores, Secretaría General de la Presidencía de la República, Correspondencia Presidential (AHM SGPR CP) Feb.11-20, 1929, Reyes Delboso to Juan Vicente Gómez, 14.2.29. Delboso appealed to Gómez to change the law.
31. FO 371/17619 Cadman to Sir R. Vansittart, 23.3.34.
32. *Ibid.* Enclosure, 'Note regarding Anglo-Persian Oil Co. Ltd. for the information of His Excellency the Venezuelan Minister in London', 22.3.34.
33. FO 371/8493 G. Armstrong to Geddes (A.), 16.12.22.
34. AHM SGPR Correspondencia del Secretario General (CS), Marzo 16-31, 1934, Diógenes Escalante to Gómez, 26.3.34.
35. FO 371/18784 Des.8 E. Keeling to Sir John Simon, 21.1.35.
36. FO 371/18783 Des.38 Keeling to Simon, 20.5.35.
37. CAB 2/8 Committee of Imperial Defence, 'Minutes of the 23 Meeting', 22.2.34.
38. *Ibid.*
39. cf. Issawi & Yeganeh, *The Economics of Middle Eastern Oil;* and Elizabeth Monroe, 'The Shaikdom of Kuwait', *International Affairs* (July 1954), 271-84.
40. In 1939 it concluded a supplementary agreement with the government, extending its concession to 80,000 sq. miles.
41. For example, the Damnan oilfield was discovered in March 1938.
42. FO 371/19845 Des.92 J. MacGregor to Eden, 24.12.35. General Velasco was the Governor of Caracas.
43. FO 371/19782 'Extract from letter written to N. Carr, Official of the Caribbean Petroleum Company in Caracas, from a friend giving account of events at Lagunillas during recent disturbances', enclosed in Des.11 MacGregor to Eden, 24.1.36.
44. FO 371/19845 MacGregor to Eden, 6.1.36.
45. FO 371/19845 Colonel Seagrin, 'Report on situation in Maracaibo', April 1936.
46. FO 371/19845 Des.11 MacGregor to Eden, 24.1.36.
47. *Oil News,* 40:1239 (Aug.27, 1936).
48. FO 371/19846 Mr. Starling to Mr. J.M. Troutbeck, 'The Petroleum Industry of Venezuela and the position of British Interests – Memorandum by the Petroleum Department', 30.3.36.
49. *Ibid.*
50. FO 371/17619 Des.23 Keeling, 'Annual Report – Venezuela, 1933', 3.3.34.
51. FO 371/19846 Des.87 MacGregor to Eden, 4.9.36.
52. The 'Shell' Transport & Trading Co., *Annual Report,* 1937.
53. *Ibid.*

54. FO 371/19846 F. Godber to F.C. Starling, 11.9.36.
55. cf. Juan Bautista Fuenmayor, *1928–1948. Veinte años de Política* (Madrid: Editorial Mediterráneo, 1968), pp.163-75.
56. Venezuela, Corte Federal y de Casación, *Memoria 1938*, Sentencia 12.
57. Lieuwen, *Petroleum in Venezuela*, p.75.
58. CAB 50/7/Secret/O.B.294 Committee of Imperial Defence, Oil Board, 'Thirteenth Annual Report', 24.1.39.
59. CAB 50/6/O.B.247 Committee of Imperial Defence, Oil Board, 'The Expropriation by the Mexican Government of the Properties of the Oil Companies in Mexico. Memorandum by the Petroleum Department', 8.4.38.
60. *Ibid.*
61. *Ibid.*
62. Lloyd C. Gardner, *Economic aspects of New Deal Diplomacy* (Madison: The University of Wisconsin Press, 1964), p.110.
63. Charlton Ogden to C. Hull & F.D. Roosevelt, 14.6.38, Roosevelt MSS President's Personal File No.3794, in Gardner, *Economic aspects of New Deal Diplomacy*, p.113.
64. CAB 50/6/O.B.247 Committee of Imperial Defence, Oil Board, 'The Expropriation by the Mexican Government of the Properties of the Oil Companies in Mexico. Memorandum by the Petroleum Department', 8.4.38.
65. *Ibid.*
66. *Ibid.*
67. CAB 50/7/Secret O.B.252 Earl de la Warr, President, Committee of Imperial Defence, Oil Board, 'Expropriation of the Properties of the Oil Companies in Mexico. Note by the Oil Board', 9.5.38.
68. CAB 2/7 Committee of Imperial Defence, 'Minutes of the 322 Meeting', 12.5.38.
69. *Ibid.*
70. Under the agreement the U.S. would buy a certain amount of silver from Mexico at prices above those prevailing on world markets.
71. Gardner, *Economic aspects of New Deal Diplomacy*, p.116.
72. CAB 50/7/Secret/O.B.294 Committee of Imperial Defence, Oil Board, 'Thirteenth Annual Report', 24.1.39.
73. FO 371/22850 Godber, 'Memorandum. Political situation in Venezuela', 22.6.38.
74. cf. Alejandro Pietri, *Lago Petroleum Corporation, Standard Oil Company of Venezuela y Compañía de Petróleo Lago contra la nación, por la negativa de exoneración de derechos de importación* (Caracas: Lit. y Tip.del Comercio, 1940).
75. FO 371/22850 'Memorandum to the Petroleum Department', Undated.
76. FO 371/24270 Des.82 D. St. Clair Gainer to Viscount Halifax, 16.9.40.
77. Venezuela, Corte Federal y de Casación, *Memoria 1938*. Sentencia 12. The government's claim of Bs. 11 million in interest charges was dismissed by the Court.
78. Cf. Gardner, *Economic aspects of New Deal Diplomacy*.
79. Archivo del Ministerio de Fomento, 'Perspectiva de la Industria Petrolera en 1940', undated.
80. Bryce Wood, *The Making of the Good Neighbor Policy* (New York: Columbia University Press, 1961), p.265.
81. *Ibid.*, p.264.
82. FO 371/22850 Des.1249E Lord Lothian to Foreign Office, 10.11.39; and D.L.T. Knudson, 'Petroleum, Venezuela, and the United States: 1920-1941' (PhD.Diss., Michigan State University, 1975).
83. According to Knudson the agreement increased the monopoly 'position of

major oil companies in the US petroleum industry'. (Knudson, 'Petroleum, Venezuela', p.234).

84. As a result of the Abyssinian Crisis the U.S. Congress in 1935 enacted the Neutrality Act which prohibited *inter alia* the export of oil for war purposes.
85. Committee of Imperial Defence, 'Sub-Committee on Oil from Coal'.
86. *Ibid.*, p.475.
87. *Ibid.*
88. *Ibid.*, p.496.
89. Cf. Payton-Smith, *Oil.*

Conclusion

The attempt to secure Britain's oil independence failed dismally. The Persian and Iraqi oilfields, expected to spearhead the new initiative, did not live up to expectations, mainly for political reasons which delayed the construction of the infrastructure needed to export the oil. For example, Iraq started to export oil only in 1934 when the pipeline to the Mediterranean was completed. The U.S., until it was displaced by Venezuela in the late 1920s, still remained Britain's largest oil supplier, and it became increasingly apparent that British policy decisions would have to take American oil interests into account. After World War I, the U.S. emerged with greater prestige demanding and succeeding in getting the world to recognise her as an emerging world power. Up to then, only the Caribbean and Central America (and to varying degrees the rest of Latin America) were recognised as American spheres of influence. The U.S. now flexed her muscles to warn the European Powers that in the future American policies would have to be taken into account, and it was over oil matters in the Middle East that this policy was carried out in its most aggressive form.

The dispute which followed after the San Remo agreement between the U.S. and Britain over the exploitation of oil resources in Mesopotamia centred on three questions: first, whether the U.S. could share in the victor's spoils when she had not declared war on Turkey; secondly, whether U.S. citizens had any claims on *TPC*'s concession; and, thirdly, whether the San Remo agreement barred American companies from operating in Iraq. American companies were eventually allowed in to develop, together with British concerns, the oil resources of the region. However, the expected oil bonanza did not materialise, with the Middle East in 1939 accounting for only 5 per cent of total world production. By far the most important source of crude oil supplies for Britain and Europe was Venezuela. *Shell, Exxon,* and to a lesser extent *Gulf Oil* supplied Europe from their Venezuelan production in 1939 with 298,000 b/d in addition to a further 188,000 b/d exported to the U.S.,[1] compared to 171,000[2] b/d which reached Europe from Iran and Iraq.

On the eve of World War II Britain depended essentially on *Shell* and *Exxon* for her oil supplies, which came mainly from Venezuela where production was controlled by the same two companies which controlled Britain's oil supplies during and after World War I. After the end of that conflict *Anglo-Persian* evolved as a major oil company, but despite the British government's 51 per cent shareholding majority, the company could not manage to produce sufficient oil to lessen Britain's dependence on foreign oil. The changing structure of the international oil industry contributed to this situation, with the 1928 Pool Division Agreement and the Red Line Agreement of the same year ensuring that the Persian and Iraqi oilfields were developed according to the needs of the major oil companies. Venezuela's development was also a result of this changing structure as in the early 1920s the companies looked for a cheap and geographically convenient source of oil. The conclusion must be drawn that Britain's post-World War I policy of lessening her dependency on foreign oil companies and American oil supplies was thwarted by the lack of oilbearing lands in the Empire and by the commercial interests of the major oil companies which controlled the international oil trade.

NOTES

1. Vernon Herbert Grigg, 'The International Price Structure of Crude Oil' (PhD.Diss., Massachussetts Institute of Technology, 1954).
2. U.S. Senate, 'Investigation of Petroleum Resources in relation to the National Welfare', Final Report of the Special Committee Investigating Petroleum Resources, *Senate Report No. 9*, 80 Cong., 1 Ses., 1947, p.35.

BIBLIOGRAPHY

1. ARCHIVES

A. United Kingdom

Public Record Office

Admiralty
  ADM 1 Admiralty Department Class, 1905–16.
  ADM 137 1914–18 War Histories.

Board of Trade
  BT 11 Commercial Dept., Correspondence & Papers.
  BT 59 Dept. of Overseas Trade, Overseas Trade & Development
    Council, 1930–39.
  BT 60 Dept. of Overseas Trade, Correspondence & Papers, 1918–39.
  BT 62 Finance Dept., Controller of Trading Accounts, Correspon-
    dence & Papers, 1918–30.

Cabinet Office
  CAB 1 Miscellaneous Records.
  CAB 2 Committee of Imperial Defence, Minutes.
  CAB 21 Registered Files.
  CAB 24 Memoranda.
  CAB 27 Committee, General Series.
  CAB 37 Cabinet Papers, Pre-1916.
  CAB 50 Committee of Imperial Defence, Oil Board, 1925–39.

Foreign Office
  FO 371 Venezuela, General Correspondence, Political, 1908–39.

Ministry of Munitions
  MUN 5 Historical Records, 1915–22.

Ministry of Power
  POWE 33 Petroleum Division, Correspondence & Papers, 1917–39.
  POWE 34 Oil Control Board, Papers, 1920–39.

Science Library (London)

S. Pearson & Sons Ltd. Archives
  Box A Files 5–7.
  Box C Files 25, 30.

B. United States of America

Records of the Department of State relating to the Internal Affairs of Venezuela, 1910-29, National Film Archives Microcopy No. 366, Reels 24-28.

C. Venezuela

Ministerio de Energía y Minas

Archivo dependiente de la División de Conservación de la Oficina Técnica de Hidrocarburos.
Archivo del Ministerio de Fomento - Unclassified.

Palacio de Miraflores

Archivo Histórico de Miraflores, Presidential & Secretary General's Correspondence, 1908-39.

2. CONTEMPORARY OFFICIAL PUBLICATIONS:

A. United Kingdom

Board of Trade Journal & Commercial Gazette, 1908-39.
Journal of the House of Commons, 1900-39.
Parliamentary Commons Debates (Hansard), 1900-39.
Public General Acts, 1865-1939.

B. United States of America

Congressional Record, Proceedings & Debates, 1900-35.
State Department, *Papers relating to the Foreign Relations of the United States,* 1900-39.

C. Venezuela

Corte Federal y de Casación, *Memoria,* 1909-39.
Ministerio de Fomento, *Memoria,* 1909-39.
Ministerio de Relaciones Exteriores, *Memoria,* 1909-39.

3. OTHER CONTEMPORARY SOURCES

Oil Facts & Figures (London, 1912-36).
Oil News (London, 1913-36).
Petroleum (London, 1939-41).
Petroleum Press Service (The Hague, 1934-40).
Petroleum Times (London, 1919-36).
Petroleum World (London, 1915-31).
The Pipeline (London), vol. 1-14.
Royal Dutch Co., *Annual Reports,* (1907-36).
The 'Shell' Transport & Trading Co. Ltd., *Annual Reports* (1907-36).
Walter R. Skinner, *The Oil & Petroleum Manual* (1910-36).
The South American Journal (London, 1900-36).
World Petroleum (1930-1).

4. UNPUBLISHED SOURCES

Edith Myretta James Blendon, 'Venezuela and the United States, 1928-1948. The impact of Venezuelan Nationalism' (PhD.Diss., University of Maryland, 1971).

Vernon Herbert Grigg, 'The International Price Structure of Crude Oil' (PhD.Diss., Massachusetts Institue of Technology, 1954).

M.R. Kent, 'British Government interest in Middle East Oil Concessions, 1900-1925' (PhD.Diss., The University of London (LSE), 1968).

David Lawrence Taylor Knudson, 'Petroleum, Venezuela, and the United States: 1920-1941' (PhD. Diss., Michigan State University, 1975).

B.S. McBeth, 'Juan Vicente Gómez and the Venezuelan Oil Industry; with special reference to the British Oil Companies' (B.Phil. Diss., Oxford University (Trinity College), 1975).

Ibid., 'Juan Vicente Gómez and the Oil Companies' (D.Phil. Diss., Oxford University (St. Antony's College), 1980).

Gholan Reza Nikpay, 'The political aspects of foreign oil interests in Iran down to 1947' (PhD. Diss., The University of London, 1955).

Arthur D. Redfield, 'Our Petroleum Diplomacy in Latin America' (PhD. Diss., The American University (Washington), 1942).

Peter Seaborne Smith, 'Petroleum in Brazil: A study in economic nationalism' (PhD. Diss., The University of New Mexico, 1969).

Gregory Frye Treverton, 'Politics and Petroleum: The International Petroleum Company in Peru' (B.A. Diss., Princeton University, 1969).

Billy Hughel Wilkins, 'The effects on the economy of Venezuela of actions by the international petroleum industry and United States regulating agencies' (PhD. Diss., The University of Texas, 1962).

5. OFFICIAL PUBLICATIONS

A. United Kingdom

'Anglo-Persian Oil Company (Acquisition of Capital). A Bill No.345, 28 July 1914', *PP* 1914, Vol. 1, 141-4.
'Anglo-Persian Oil Company (Acquisition of Capital) Amendment. Bill 239, 11.12.19', *PP* 1919, Vol. 1, 15-8.
'Agreement with Anglo-Persian Oil Company, with an explanatory Memorandum and the report of the Commission of Experts on their local investigation', *PP* 54 (1914), Cmd. 7419, 505-39.
Board of Trade, 'Report of National Fuel and Power Committee', *PP* 1928-9, Vol. 6, Cmd. 3201, 513-50.
Ibid., 'Second report of National Fuel and Power Committee', *PP* 1928-9, Vol. 6, Cmd. 3252, 551-98.
'British Hydrocarbon Oils Production. A Bill (Bill 47), 21.12.33', *PP* 1933-4, Vol. 1, 143-6.
Committee on Imperial Defence, 'Sub-Committee on Oil from Coal. Report', *PP* 1937-38, Vol. 12, Cmd. 5665, 439-512.
'Convention between His Majesty and His Majesty the King of Iraq and the President of the United States of America regarding the rights of the United States and of its nationals in Iraq, with Protocol and Exchange Notes, London, 9 January 1930', *PP* 1930-31, Vol. 34, Cmd. 3833, 23-76.
'Correspondence between His Majesty's Government and the United States Ambassador respecting economic rights in Mandated Territories (Misc. No. 10)', *PP* 1921, Vo. 43, Cmd. 1226, 481-94.
'Despatch to His Majesty's Ambassador at Washington, enclosing a Memorandum on the Petroleum situation (Misc. No. 17 (1921))', *PP* 1921, Vo. 43, Cmd. 1351, 495-500.
'Franco-British Convention of 23 December 1920 on certain points connected with the Mandates for Syria and the Lebanon, Palestine and Mesopotamia (Misc. 4 (1921))', *PP* 1921, Vol. 42, Cmd. 1195, 669-72.
H.M. Petroleum Executive, 'Report of the Inter-Departmental Committee on the Employment of Gas as a source of power, especially in motor vehicles, in substitution for petrol and petroleum products', *PP* 1919, Vol. 22, Cmd. 263, 521-68.
Ibid., 'Report of the Inter-Departmental Committee on various matters concerning the production and utilization of alcohol for power and traction purposes', *PP*. 1919 Vol. 10, Cmd. 218, 117-24.
H.M. Petroleum Institute, *Petroleum* (London: John Murray (Imperial Institute Monographs on Mineral Resources, with special reference to the British Empire), 1921).
Imperial Conference, 1930, 'Summary of Proceedings', *PP* 1930-31, Vol. 14, Cmd. 3717, 569-700.
Ibid., 'Appendices to the Summary of Proceedings', *PP* 1930-31, Vol. 14, Cmd. 3718, 701-972.

Imperial War Conference, 1918, 'Extracts from Minutes of Proceedings and Papers laid before the Conference', *PP* 1918, Vol. 16, Cmd. 9177, 691-942.

'Lausanne Conference on Near Eastern Affairs, 1922-23. Records of Proceedings and Draft of Terms of Peace (Turkey) No. 1', *PP* 1923, Vol. 26, Cmd. 1814, 1-861.

'League of Nations. Decision relating to the Turco-Irak frontier adopted by the Council of the League of Nations, Geneva, 16 December 1925 (Misc. 17 (1925))', *PP* 1924-5, Vol. 31, Cmd. 2562, 549-54.

'League of Nations. Letter from His Majesty's Government to the Secretary General of the League of Nations and Proceedings of the Council of the League regarding the determination of the Turco-Irak frontier and the application to Irak of the Provisions of article 22 of the Covenant of the League (Misc. 3 (1926))', *PP* 1926, Vol. 30, Cmd. 2624, 693-712.

'League of Nations. Report by M. Unden on the question of the Turco-Irak frontier, Geneva, 16 December 1925 (Misc. 16, 1925)', *PP* 1924-5, Vol. 31, Cmd. 2565, 563-72.

'League of Nations. Report to the League of Nations by General F. Laidoner on the situation in the locality of the provisional line of the frontier between Turkey and Irak fixed at Brussels on 29 October 1923. Mosul, 23 November 1925 (Misc. 15 (1925))', *PP* 1924-5, Vol. 31, Cmd. 2557, 541-8.

'League of Nations. Thirty-Ninth Session of the Council. Report by the Rt. Hon. Sir Austen Chamberlain K.G. M.P. (Misc. 4 (1926))', *PP* 1926, Vol. 30, Cmd. 2646, 637-42.

'Licence to drill for Petroleum granted by the Secretary of State for Mines to His Grace the Duke of Devonshire', *PP* 1923, Vol. 19, Cmd. 1873, 659-66.

'Memorandum of Agreement between M. Philippe Berthelot, Directeur des Affaires Politiques et Commerciales au Ministère des Affaires Etrangères and Professor Sir John Cadman K.C. M.G., Director in Charge of His Majesty's Petroleum Department', *PP* 1920, Vol. LI, Cmd. 675, 895-8.

Ministry of Munitions, 'Copy of an agreement made between the Minister of Munitions and S. Pearson & Sons Limited on 10 September 1918 for management of Petroleum Development', *PP* 1918, Vo. XV, Cmd. 9188, 743-50.

Ministry of Munitions of War, 'Report of a Committee appointed by the Rt. Hon. The Minister of Munitions respecting the production of fuel oil from home sources', *PP* 1918, Vol. X, Cmd. 9128, 515-22.

'Petroleum Production (Licences). Licence dated 12 May 1919, granted by the Minister of Munitions to Oilfields of England, Limited', *PP* 1919, Vol. XLII, Cmd. 1917, 1047-52.

'Policy in Iraq. Memorandum by the Secretary of State for the Colonies', *PP* 1929-30, Vol. 25, Cmd. 3440, pp.199-202.

'Profiteering Acts, 1919 & 1920. Second Report on Motor Fuel prepared by a Sub-Committee appointed by the Standing Committee on the

Investigation of Prices', *PP* 1921, Vol. 6, Cmd. 1119, 793-804.

'Protocol between the Government of the United Kingdom, France and Iraq for the transfer from the United Kingdom to Iraq of certain rights and obligations under the San Remo oil agreement of 24 April 1920, and the Convention between the United Kingdom and France of 23 December 1920 relating to Mandates in the Middle East, Geneva, 10 October 1932 (Treaty Series No. 37 (1932))', *PP* 1932-3, Vol. 27, Cmd. 4220, 401-2.

'Report on Motor Fuel prepared by a Sub-Committee appointed by the Standing Committee on Investigation of Prices', *PP* 1919, Vol. 23, Cmd. 597, 573-82.

'Rt. Hon. William C. Bridgeman M.P. (Secretary for Mines) and Oilfields of England Limited License', *PP* 1921, Vol. 31, Cmd. 1434, 541-6.

Royal Commission on the Natural Resources, Trade, and Legislation of Certain Portions of His Majesty's Dominions, Royal Commission Dominions, 'Memorandum and tables relating to the Food and Raw Materials requirements of the United Kingdom', *PP* 1914–16, Vol. XIV, Cmd. 8123 (1915), 371-98.

Ibid., 'Final report of the Royal Commission on the Natural Resources, Trade and Legislation of Certain Portions of His Majesty's Dominions', *PP* 1917, Vol. X, Cmd. 8462, pp. 1-205.

'The Minister of Munitions and Reginald Gilbey. Copy. Licence as to boring for petroleum at Weston, in the county of Leicester', *PP* 1920, Vol. 30, Cmd. 1031, 257-62.

'Treaty of Peace with Turkey, and other instruments signed at Lausanne on 24 July 1923, together with agreements between Greece and Turkey signed on 30 January 1923, and subsidiary documents forming part of the Turkish Peace Settlement (Treaty Series No. 16 (1923))', *PP* 1923, Vol. 25, Cmd. 1929, 533-784.

B. United States of America

U.S. Congress, *Petroleum Industry,* Investigation of Economic Power, Hearings before the Temporary National Economic Committee, Part 14, Section 1, Part 14-A, Section 4, Part 15 & 17, 76 Cong., 1 Ses., 1940.

Ibid., Foreign Contracts Act, Joint Hearings before a Sub-Committee of the Committee on the Judiciary U.S. Senate and the Special Committee Investigating Petroleum Resources, 79 Cong., 1 Ses., 1945.

U.S Dept. of Interior, *Report on the cost of Producing Crude Petroleum* (Washington: USGPO 1936).

U.S. Dept. of Trade, 'British Petroleum Trade in 1925', Bureau of Foreign and Domestic Commerce, *Trade Information Bulletin,* No. 407, April 1926.

U.S. Federal Trade Commission, 'Advance in the Price of Petroleum Products', *House Report 801,* 66 Cong., 2 Ses., 1920.

Ibid., Report of the Federal Trade Commission on the Pacific Coast petroleum Industry (Washington: USGPO 1921).

Ibid., Panhandle Crude Petroleum (Washington: USGPO 1928).

U.S. Federal Trade Commission, *The International Petroleum Cartel*, Staff Report to the Federal Trade Commission Submitted to the Sub-Committee on Monopoly of the Select Committee on Small Business, United States Senate, Committee Print No. 6, 82 Cong., 2 Ses., 1952.

U.S. Federal Trade Commission, *Rates of return (after taxes) for 516 identical companies in 25 selected manufacturing industries, 1940, 1947-52* (Washington: USGPO 1954).

John W. Frey & H. Chandler Ide, *A History of the Petroleum Administration for War, 1941-45* (Washington: USGPO 1946).

H.A. Garfield, *Final report of the U.S. Fuel Administration 1917-1919* (Washington: USGPO 1919).

U.S. House of Representatives, 'Regulating importation of petroleum and related products', Hearings before the Committee on Ways & Means, 71 Cong., 3 Ses., 1931.

Ibid., 'Production costs of crude petroleum and of refined petroleum products', *House Doc. No. 195*, 72 Cong., 1 Ses., 1932.

Ibid., 'Conservation of Petroleum', Hearings before the Committee on Ways & Means, 73 Cong., 1 Ses., 1933.

Ibid., 'Crude Petroleum', Hearings before the Committee on Interstate and Foreign Commerce, 73 Cong., 1 Ses., 1933.

Ibid., 'Petroleum Investigation', Hearings before a Sub-Committee of the Committee on Interstate and Foreign Commerce, 73 Cong. (Recess), 1934.

Ibid., 'Effects of foreign imports on independent domestic producers', *House Report 2344*, 81 Cong., 2 Ses., 1950.

Joseph E. Pogue, *Prices of petroleum and its products during the war* (Washington: USGPO 1919).

U.S. Senate, 'Restrictions on American Petroleum Prospectors in certain foreign countries', *Senate Doc. 11*, 67 Cong., 1 Ses., 1921.

Ibid., 'Oil prospecting in foreign countries', *Senate Doc. 39*, 67 Cong., 1 Ses., 1921.

Ibid., 'Diplomatic correspondence with Colombia in connection with the treaty of 1914 and certain oil concessions', *Senate Doc. 64*, 68 Cong., 1 Ses., 1924.

Ibid., 'Oil concessions in foreign countries', *Senate Doc. 97*, 68 Cong., 1 Ses., 1924.

Ibid., 'Regulating importation of petroleum and regulated products', Hearings before the Committee on Commerce, 71 Cong., 3 Ses., 1931.

Ibid., 'Regulating imports of petroleum', *Senate Report 1476*, 71 Cong., 3 Ses., 1931.

Ibid., 'Cost of crude petroleum in 1931', *Senate Doc. 267*, 71 Cong., 3 Ses., 1931.

Ibid., 'Patents', Hearings before the Committee on Patents, 77 Cong., 2 Ses., 1942.

Ibid., 'American petroleum interests in foreign countries', Hearings before a Special Committee Investigating Petroleum Resources, 79 Cong., 1 Ses., 1946.

Ibid., 'Foreign Contracts Act', Joint hearings before a Sub-Committee of the Committee of the Judiciary and the Special Committee Investigating Petroleum Resources, 79 Cong., 1 Ses., 1945.

Ibid., 'Investigation of Petroleum Resources', Hearings before a Special Committee Investigating petroleum resources, 79 Cong., 1 Ses., 1946.

Ibid., 'Wartime petroleum policy under the petroleum Administration for War', Hearings before a Special Committee Investigating Petroleum Resources, 79 Cong., 1 Ses., 1945.

Ibid., 'Diplomatic protection of American petroleum interests in Mesopotamia, Netherlands East Indies and Mexico', *Senate Doc. 43*, 79 Cong., 1 Ses., 1945.

Ibid., 'Investigation of petroleum resources in relation to the National Welfare', Final Report of the Special Committee Investigating Petroleum Resources, *Senate Report No. 9*, 80 Cong., 1 Ses., 1947.

U.S. Tariff Commission, 'Petroleum', War Changes in Industry Series No. 17, 1946.

John K. Towles, 'Petroleum trade and industry in the United Kingdom', Bureau of Foreign and Domestic Commerce, Supplement to Commerce Reports, *Trade Promotion Series No. 80*, 29. Jan. 23.

C. Venezuela

Venezuela, Ministerio de Minas e Hidrocarburos, *Petróleos crudos de Venezuela y otros paises* (Caracas: Ministerio de Minas e Hidrocarburos, 1959) 2nd ed.

Ibid., Petróleo y otros datos estadísticos (Caracas: 1964).

6. SECONDARY SOURCES

A. Books

M.A. Adelman, *The World Petroleum Market* (Baltimore: The Johns Hopkins University Press, 1972).

Francisco Alonso González, *Historia y Petróleo. México: El problema del petróleo* (Madrid: Editorial Ayuso, 1972).

Ralph Arnold et al., *The First Big Oil Hunt: Venezuela 1911–1916* (New York: Vantage Press, 1960).

Raymond Foss Bacon & William Allen Manor, *The American Petroleum Industry* (New York: McGraw-Hill Book Co. Inc., 1916) 2 vols.

Sir Boverton Redwood Bart, *A Treatise on Petroleum* (London: Charles Griffin & Co., 1922) 4th ed. 3 vols.

J. Leonard Bates, *The Origins of Teapot Dome. Progressive Parties and Petroleum 1900–1921* (Urbana: University of Illinois Press, 1963).

Kendall Beaton, *Enterprise in Oil. A History of Shell in the United States* (New York: Appleton-Century Crofts Inc., 1957).

Paul G. Bradley, *The Economics of Crude Petroleum Production* (Amsterdam: North-Holland Publishing Co., 1967).

Benjamin T. Brooks, *Peace, Plenty and Petroleum* (Lancaster: The Jacques Cattell Press, 1944).

Amado Canelas, *Petróleo, Imperialismo y Nacionalismo* (La Paz: Libreria 'Altiplano', 1963).

Ralph Cassady jr., *Price Making and Price Behaviour in the Petroleum Industry* (New Haven: Yale University Press, Petroleum Monograph Series No. 1, 1954).

Roy C. Cook, *Control of the Petroleum Industry by Major Oil Companies* (Washington: USGPO, 1941).

David T. Day ed., *A Handbook of the Petroleum Industry* (New York: John Wiley & Sons Inc., 1922) 2 vols.

E.H. (Nicholas) Davenport & S.R. Cooke, *The Oil Trusts and Anglo-American Relations* (London: Macmillan & Co., 1923).

Melvin de Chazeau & Alfred E. Kahn, *Integration and Competition in the Petroleum Industry* (Port Washington (New York): Kennikat Press, 1973).

Francis Delaisi, *Oil, its influence in Politics* (London: The Labour Publishing Co. Ltd. & George Allen & Unwin, 1922).

Ludwell Denny, *We fight for Oil* (New York: Alfred A. Knopf, 1928).

Ibid., America Conquers Britain (New York: Alfred A. Knopf, 1930).

Sir Henri Deterding, *An International Oilman* (London: Ivor Nicholson & Watson Ltd., 1934).

L. Domeratsky *et al., Regulations of economic activities in foreign countries* (Washington: USGPO, 1941).

William Diebold jr., *New Directions in our Trade Policy* (New York: Council on Foreign Relations, 1941).

Robert W. Dunn, *American Foreign Investments* (New York: B.W. Huebsch & the Viking Press, 1926).

Edward Mead Earle, *Turkey, the Great Powers and the Baghdad Railway* (London: Macmillan & Co. Ltd., 1923).

Editors of *Look, Oil for Victory. The Story of Petroleum in War and Peace* (New York: McGraw-Hill Book Co. Inc., 1946).

Brooks Emeny, *The Strategy of Raw Materials. A Study of America in Peace and War* (New York: The Macmillan Co., 1954).

Robert Engler, *The Politics of Oil – a Study of Private Power and Democratic Directions* (New York: The Macmillan Co., 1961).

Leonard M. Fanning, *American Oil Operations Abroad* (New York: McGraw-Hill Book Co. Inc., 1947).

Nasrollah Saipour Fatemi, *Oil Diplomacy. Powderkeg in Iran* (New York: Whittier Books Inc., 1954).

Herbert Feis, *As Seen from E.A. Three international Episodes* (New York: Alfred A. Knopf, 1947).

Louis Fischer, *Oil Imperialism* (London: George Allen & Unwin, 1926).

Helmut J. Frank, *Crude Oil Prices in the Middle East – A Study of Oligopolistic Price Behaviour* (New York: Frederick A. Praeger Publishers, 1966).

Paul H. Frankel, *Oil: The Facts of Life* (London: Weidenfeld & Nicolson, 1962).

Ibid., *Essentials of Petroleum* (London: Frank Cass & Co., 1969).

Arturo Frondizi, *Petróleo y Política* (Buenos Aires: Editorial Raigal, 1956) 2nd ed.

Juan Bautista Fuenmayor, *1928-1948: Veinte años de política (Madrid: Editorial Mediterráneo, 1968).*

Jaime Galarza Zavala, *El festín del petróleo* (Quito: Ediciones Solitierra, 1972) 2nd ed.

Lloyd C. Gardner, *Economic Aspects of New Deal Diplomacy* (Madison: The University of Wisconsin Press, 1964).

F.C. Gerretson, *History of the Royal Dutch* (Leiden: E.J. Brill, 1957) 4 vols.

George S. Gibb & E.H. Knowlton, *The Resurgent Years, 1911-1927* (New York: Harper & Bros., 1956).

P.H. Giddens, *Standard Oil Company (Indiana)* (New York: Appleton Century Crofts Inc., 1955).

Daniel C. Hamilton, *Competition in Oil. The Gulf Coast Refinery Market, 1925-1950* (Cambridge (Mass.): Harvard University Press, 1958).

Frank G. Hanighen, *The Secret War* (New York: John Day Co., 1934).

J.E. Hartshorn, *Politics and World Oil Economics* (New York: Frederick A. Praeger Publishers, 1962).

A. Eugene Havens & Michel Romieux, *Barrancabermeja. Conflictos sociales en torno a un centro petrolero* (Bogotá: Universidad Nacional, 1966).

J.D. Henry, *Oil, Fuel and the Empire* (London: Bradbury, Agnew & Co., 1908).

Robert Henriques, *Marcus Samuel, First Viscount Bearsted and founder of the 'Shell' Transport and Trading Company, 1853-1927* (London: Barrie & Rockliff, 1966).

Ralph Hewins, *Mr. Five Per cent, the Biography of Calouste Gulbenkian* (London: 1957).

Ervin Hexner, *International Cartels* (London: Sir Isaac Pitman & Sons Ltd., 1946).

Charles Issawi & Mohammed Yeganeh, *The Economics of Middle Eastern Oil* (London: Faber & Faber, 1963).

John Ise, *The United States Oil Policy* (New Haven: Yale University Press, 1926).

V.A. Kalichevsky, *The Amazing Petroleum Industry* (New York: Reinhold Publishing Corp., 1943).

George P. Kerr, *Time's Forelock. A Record of Shell's Contribution to Aviation in the Second World War* (London: The Shell Petroleum Co., 1948).

R.J. Kirwin *Economic Conditions in Venezuela,* Board of Trade, Department of Overseas Trade Series (London: HMSO, 1932).

Luis Laurie Solis, *La diplomacia del petróleo y el caso de 'La Brea y Pariñas'* (Lima: Editorial Minema, 1934).

Wayne A. Leeman, *The Price of Middle East Oil. An Essay in Political Economy* (New York: Cornell University Press, 1962).

C.K. Leith, *World Minerals and World Politics. A Study of Minerals in their*

Political and International Relations (New York: McGraw-Hill Book Co. Inc., 1931).

George Lenczowski, *Oil and State in the Middle East* (Ithaca: Cornell University Press, 1960).

Pierre L'Espagnol de la Tramerye, *The World Struggle for Oil* (London: George Allen & Unwin , 1923) 3rd ed.

Walter J. Levy, *The Past, Present and Likely Future Price Structure for the International Oil Trade* (The Hague; Leiden: E.J. Brill, 1951).

Edwin Lieuwen, *Petroleum in Venezuela* (Berkeley: University of California Press, 1954).

Ernest Raymond Lilley, *The Oil Industry* (London: Constable & Co. Ltd., 1926).

Stephen Hemsley Longrigg, *Oil in the Middle East. Its Discovery and Development* (London: Oxford University Press, 1968) 3rd ed.

H. Longhurst, *Adventure in Oil* (London: Sidgwick & Jackson, 1959).

Pedro N. López, *Política petrolera* (La Paz: Imprenta Boliviana, 1929).

McBeth, B.S., *Royal Dutch-Shell U.S. Venezuela* (Oxford: Oxford Microform, 1982).

McBeth, B.S., *Juan Vicente Gómez and the Oil Companies in Venezuela, 1908-1935* (Cambridge: Cambridge University Press, 1983).

John G. McClean & Robert W. Haigh, *The Growth of the Integrated Oil Companies* (Norwood (Mass.): Plimpton Press, 1954).

Lorenzo Meyer, *México y Estados Unidos en el conflicto petrolero, 1917-1942* (México: El Colegio de Mexico, 1968).

Raymond F. Mikesell & Hollis B. Chenery, *Arabian Oil* (Chapel Hill: The University of North Carolina Press, 1949).

Zuhayr Mikdashi, *A Financial Analysis of Middle Eastern Oil Concessions, 1901-65* (New York: Frederick A. Praeger Publishers, 1966).

Arthur C. Millspaugh, *Americans in Persia* (Washington: The Brookings Institution, 1946).

Anton Mohr, *The Oil War* (London: Martin Hopkinson Co. Ltd., 1926).

Parker Thomas Moon, *Imperialism and World Politics* (New York: The Macmillan Co., 1926).

Enrique Mosconi, *El petróleo en Argentina 1922-1930 y la ruptura de los trusts petrolíferos ingleses y norteamericanos en el 1 de agosto de 1929* (Buenos Aires: Talleres Gráficos Ferrari Hnos., 1936).

Gerald D. Nash, *The United States Oil Policy, 1890-1964* (Pittsburgh: University of Pittsburgh Press, 1968).

Burl Noggle, *Teapot Dome. Oil and Politics in the 1920's* (New York: W.W. Norton & Co. Inc., 1965).

Edwin G. Nourse, *America's Capacity to Produce* (Washington: The Brookings Institution , 1934).

Richard O'Connor, *The Oil Barons, Men of Greed and Grandeur* (London: Hart-Davis MacGibbon Ltd., 1972).

Campbell Osborn, *Oil Economics* (New York: McGraw-Hill Book Co. Inc., 1932).

R. Page Arnot, *The Politics of Oil - An Example of Imperialist Monopoly* (London: The Labour Publishing Co. Ltd., 1924).

D.J. Payton-Smith, *Oil. A Study of War-Time Policy and Administration* (London: HMSO, 1971).

Edith Penrose, *The International Firm in Developing Countries. The International Petroleum Industry* (London: George Allen & Unwin, 1968).

S.B. Pettengill, *Hot Oil* (New York: Economic Forum Co., 1936).

H. ST. J.B. Philby, *Arabian Oil Ventures* (Washington: The Middle East Institute, 1964).

Alejandro Pietri, *La Petroleum Corporation, Standard Oil Company of Venezuela, y Compañía de Petróleo Lago contra la Nación por la negativa de exoneración de derechos de importación* (Caracas: Lit. y Tip. del Comercio, 1940).

Joseph E. Pogue, *The Economics of Petroleum* (New York: John Wiley & Sons Inc., 1921).

Ibid., Prices of Petroleum and its Products during the War (Washington: USGPO, 1919).

Domingo Alberto Rangel, *Capital y Desarrollo. El Rey Petróleo,* (Caracas: UCV, 1970). 2 vols.

Merril Rippy, *Oil and the Mexican Revolution* (Leiden: E.J. Brill, 1972).

Glynn Roberts, *The Most Powerful Man in the World* (New York: Covici-Friede Publishing, 1938).

Eugene Rostow, *A National Policy for the Oil Industry* (New Haven: Yale University Press, 1948).

John Rowland and Basil, second Baron Cadman, *Ambassador for Oil. The Life of John Cadman, First Baron Cadman* (London: Herbert Jenkins, 1960).

Edward H. Shaffer, *The Oil Import Program of the United States* (New York: Frederick A. Praeger Publishers, 1968).

Ronald B. Shuman, *The Petroleum Industry. An Economic Survey* (Norman: University of Oklahoma Press, 1940).

Benjamin Shwadran, *The Middle East, Oil and the Great Powers* (New York: Council for Middle Eastern Affairs, 1959) 2nd ed.

Eugene Staley, *Raw Materials in Peace and War* (New York: Council on Foreign Relations, 1937).

G.W. Stocking, *The Oil Industry and the Competitive System* (Boston: Houghton-Mifflin, 1924).

Ibid., Middle East Oil. A Study in Politics and Economic Controversy (London: The Penguin Press (Allen Lane), 1971).

L.P. Elwell Sutton, *Persian Oil. A Study in Power Realities* (London: Lawrence & Wishart Ltd., 1955).

A.J.P. Taylor, *English History, 1914-1945* (Oxford: Oxford University press, 1965).

Christopher Tugendhat, *Oil, the Biggest Business* (London: Eyre & Spottiswoode, 1968).

K.S. Twitchell, *Saudi Arabia, with an Account of the Development of its Natural Resources* (Princeton: Princeton University Press, 1953).

Myron W. Watkins, *Oil Stabilisation or Conservation? A Case*

Study in the Organisation of Industrial Control (New York: Harper & Bros. Publishers, 1937).

Benjamin H. Williams, *Economic Foreign Policy of the United States* (New York: McGraw-Hill Book Co. Inc., 1929).

Harold F. Williamson *et al.*, *The American Petroleum Industry; The Age of Energy, 1899–1959* (Evanston: Northwestern University Press, 1963).

Bryce Wood, *The Making of the Good Neighbor Policy* (New York: Columbia University Press, 1961).

B. Articles

'Arguments presented against Venezuelan Treaty', *Oil & Gas Journal*, (18. Aug. 38), p.23 & 34.

'Arrival of American oil in Europe and Colby's note produces effect', *Oil, Paint & Drug Reporter*, (20. Dec. 20).

Robert Barnes, 'International Oil Companies confront governments: a half century of experience', *International Studies Quarterly*, 16:4 (Dec, 1972), 454-71.

J. Leonard Bates, 'The Teapot Dome Scandal and the election of 1924', *American Historical Review*, XL (Jan. 1955), 302-22.

A.C. Bedford, 'The world oil situation', *Foreign Affairs*, 1:3 (15. March 23), 96-107.

John R. Bradley, 'Fuel and power in the British Empire', U.S. Bureau of Foreign and Domestic Commerce, *Trade Promotion Series No. 161*, (1935).

Albert D. Brokaw, 'Oil', *Foreign Affairs*, 8:1 (Oct.1927), 89-105.

J.D. Butler, 'The influence of economic factors on the location of oil refineries (with primary reference to the world outside the U.S.A. and U.S.S.R)', *The Journal of Industrial Economics*, 1:3 (July 1959), 187-201.

Geoffrey Chandler, 'The myth of oil power, international groups and national sovereignty', *International Affairs*, 46:4 (Oct. 1970), 710-8.

'Conference to hear oil plans', *Oil, Paint & Drug Reporter*, (9. Nov. 21).

Paul Davidson, 'Public policy problems of the domestic crude oil industry', *American Economic Review*, 53:1 (March 1963), 85-108.

E. de Golyer, 'Petroleum in two wars', *Petroleum*, 4:4 (Aug. 1941), 88-92.

William Diebold jr., 'Oil import quotas and "Equal Treatment"', *American Economic Review*, 30:3 (Sept. 1940), 569-73.

J.B. Dirlam, 'The petroleum industry', in Walter Adams ed., *The Structure of American industry. Some Case Studies* (New York: The Macmillan Co., 1954), pp.236-73.

Edward Mead Earle, 'International financial control of raw materials'; *Proceedings of the Academy of Political Science*, 12:1 (July 1926), 188-97.

'El petróleo sintético', *Boletín de la Cámara de Comercio de Caracas*, 17:175 (1 June 28), 41002-3.

'End of Venezuela's oil workers strike', *Petroleum Press Service,* 4:5 (29 Jan. 37), pp.57.

Harry Z. Evan, 'The multinational oil company and the nation state', *Journal of World Trade and Law,* 4:5 (Oct. 1970), 666-85.

'Financial position of American oil indsutry strengthened during the Depression'. *Petroleum Press Service,* 1:18 (15. Sept. 34) 4-5.

Ray C. Gerhardt, 'Inglaterra y el petróleo mexicano durrante la Primera Guerra Mundial', *Historia Mexicana,* 25:1 (July-Sept. 1975), 118-42.

'Giant struggle for control of the world's oil supplies', *New York Times,* (27. June 20).

W.O. Henderson, 'German economic penetration in the Middle East, 1870-1914', *The Economic History Review,* 18:1-2 (1948), 54-64.

Miriam Jack, 'The purchase of the British Government's shares in the British Petroleum Company; 1912-1914', *Past & Present,* No. 39 (1968), 139-69.

Chester Lloyd Jones, 'Oil in the Caribbean and elsewhere', *North American Review,* 202 (Oct. 1915), 536-43.

Wayne A. Leeman, 'Crude oil prices in the United States at the Gulf Coast', *The Journal of Industrial Economics,* 5:3 (July 1957), 180-91.

Robert Liefman, 'International Cartels', *Harvard Business Review,* 5:2 (Jan. 1927), 129-48.

E. Mackay Edgar, 'The petroleum resources of the world', *Sperling's Journal,* (Sept. 1919).

Rene Pierce Manes, 'The quality and pricing of crude oil: The American Experience', *The Journal of Industrial Economics,* 12:2 (March 1964), 151-62.

I.A. Manning, 'Petroleum production – Colombia', U.S. Department of Commerce & Labour, Bureau of Manufacturers, *Monthly Consular & Trade Reports,* No. 332 (May 1908), 154-55.

Elizabeth Monroe, 'The Shaikdom of Kuwait', *International Affairs* (July 1954), 271-84.

'Oil a casus belli', *Journal of Commerce* (New York, 6. Aug. 26).

E.T. Penrose, 'Profit sharing between producing countries and oil companies in the Middle East', *Economic Journal,* 69:274 (June 1959), 238-254.

Ibid., 'Middle East Oil: The international distribution of profits and income taxes', *Economica,* 27:107 (Aug. 1960), 203-13.

'Producción mundial petrolera en 1920', *Boletín del Ministerio de Fomento,* 2:16 (Jan. 1922), 12.

Mark L. Requa, 'Report of the Oil Division, 1917-1919' in H.A. Gardield, *Final Report of the U.S. Fuel Administration, 1917-1919* (Washington: USGPO, 1919).

'Shifts in European supply and the Iraq oil', *Petroleum Press Service,* 1:15 (1 Aug. 1932), 1-3.

L.C. Snyder, 'The petroleum resources of the United States', *Proceedings of the Academy of Political Science,* 12:1 (July 1926), 159-167.

S.A. Swensrud, 'The relation between crude oil and production
   prices', *Bulletin of the American Association of Petroleum Geologists*, 23:6
   (1939), 765-808.
'Venezuela, development of the petroleum industry', *Board of
   Trade Journal & Commercial Gazette*, cxx: 1629 (23 Feb. 28), 254.
Arthur C. Veatch, 'Oil, Great Britain and the United States',
   Foreign Affairs, 9:4 (July 1931), 663-673.

INDEX

Printed and bound by CPI Group (UK) Ltd, Croydon, CR0 4YY

22/10/2024

01777621-0001